R W Carter

CONCEPTS OF

SYMBIOGENESIS

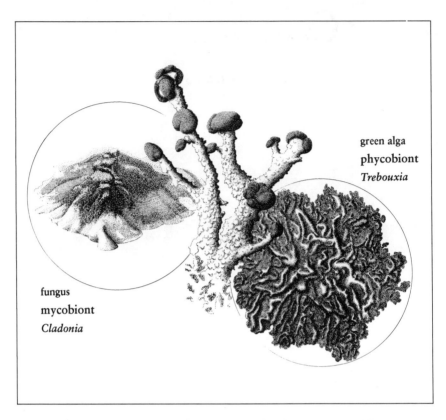

green alga
phycobiont
Trebouxia

fungus
mycobiont
Cladonia

Drawing of British soldier lichen, *Cladonia Cristatella*, by Christie Lyons, used by permission of Lynn Margulis

Concepts of Symbiogenesis

A Historical and Critical Study of the Research of Russian Botanists

LIYA NIKOLAEVNA KHAKHINA

Edited by

Lynn Margulis and Mark McMenamin

Translated by

Stephanie Merkel and Robert Coalson

With an appendix on Ivan E. Wallin by
Donna C. Mehos

A Russian and Western History of

Symbiosis as an Evolutionary Mechanism

Yale University Press

New Haven and London

Originally published in Russian as *Problema simbiogeneza: istoriko-kritichesky ocherk issledovany otechestvennykh botanikov* by L. N. Khakhina. Copyright by the Institute of the History of Natural Science and Technology, The Academy of Sciences of the USSR, Leningrad, Nauka, 1979.

Designed by Jill Breitbarth and set in Sabon and Bodoni types by Keystone Typesetting, Inc., Orwigsburg, Pennsylvania.
Printed in the United States of America by BookCrafters, Chelsea, Michigan.

Library of Congress Cataloging-in-Publication Data

Khakhina, Liîa Nikolaevna.
[Problema simbiogeneza. English]
Concepts of symbiogenesis : historical and critical study of the research of Russian botanists / Liya Nikolaevna Khakhina ; edited by Lynn Margulis and Mark McMenamin ; translated by Stephanie Merkel and Robert Coalson.
p. cm. — (Bio-origins series)
Translation of: Problema simbiogeneza.
Includes bibliographical references and index.
ISBN 0-300-04816-5 (alk. paper)
1. Symbiogenesis. 2. Biotic communities—Evolution.
I. Margulis, Lynn, 1938–
II. McMenamin, Mark A. III. Title.
IV. Series.
QH378.K4313 1992
575.01'6—dc20 92-14466 CIP

A catalogue record for this book is available from the British Library.

The paper in this book meets the guidelines for permanence and durability of the Committee on Production Guidelines for Book Longevity of the Council on Library Resources.

10 9 8 7 6 5 4 3 2 1

CONTENTS

L. N. KHAKHINA
Concepts of Symbiogenesis: A Historical and Critical
Study of the Research of Russian Botanists

F OREWORD

Although it faced challenges of harsh and penetrating criticism from many quarters, Russian evolutionary biology entered the twentieth century in robust health. Darwinian ideas had more supporters in the academic community than ever before, and Darwinian scholars dominated the university courses in general biology. University textbooks in biology, almost without exception, were built upon the ideas presented in the *Origin of Species* and the *Descent of Man*. The evolutionary approach formulated by the great English naturalist carried its authority to new domains of biological research. Russian biologists, led by A. N. Severtsov, were interested in building a bridge to connect Darwin's theoretical structures with the avalanches of new empirical data. The country had just received the first publication of Darwin's complete works, and preparations were under way for a second edition offering a larger share of Darwin's scientific legacy and fewer translation errors. In 1909, the universities and learned societies, led by the St. Petersburg Academy of Sciences, commemorated and celebrated the one-hundredth anniversary of Darwin's birth and the fiftieth anniversary of the publication of the *Origin of Species*. Overflowing crowds at numerous celebrations heard more than 150 papers that provided a magnificent display not only of the deep roots Darwin's theory had taken in Russian biology but also the triumphs of rationality over the forces of mysticism and superstition.

On entering the twentieth century, Darwinism encountered formidable challenges that originated mainly in various theoretical orientations in biology. In a state of intense fermentation, biology fed on new theoretical thought and methodological achievements, sometimes threatening to uproot the very foundations on which Darwin had built his evolutionary principles. Modern genetics, built on the resurrected theory of Gregor Mendel, made an auspicious entrance soon after 1900, and in 1909 the Institute of Applied Botany, a research component of the Ministry of Agriculture, showed strong signs of harnessing the tools of the new science in efforts to improve the existing species of cultivated plants. Neo-Lamarckism appeared in several forms, all emphasizing the environment as the primary factor of evolution and the hereditary transmission of acquired characteristics. Several variations of psychological theories considered the behavioral aspect of biological evolution. Influential authorities, led by I. P. Borodin, supported an integration of the selective principles of vitalism into general evolutionary theory. Vitalism, at this time, was undergoing a strong revival in western Europe in both idealistic metaphysics and in its opposition to mechanistic philosophy. As a branch of evolutionary biology, *vitalism* (asserting the uniqueness of living matter) was engaged in a bitter war with *mechanism,* as shown by the growing dependence of experimental biology on Newtonian physics and chemistry.

In Russia, more than in Western countries, it was eminently clear that the emergence of a critical attitude toward classical Darwinism was much stronger in botany than in zoology. Sergei I. Korzhinsky, a leading expert in plant geography, heralded the rebirth of Mendelian genetics and the triumph of mutation theory in 1899. The first Russian designs for experimental work in the genetics of cultivated cereals were prepared by R. E. Regel. A biologist turned philosopher, V. P. Karpov, advanced a brand of holistic organicism, which injected elements of vitalism into classical Darwinism. Andrei S. Famintsyn worked diligently to add a psychological dimension to Darwin's theory and deserved to be counted among the pioneers of ethology. Vladimir L. Komarov, a leading botanist, saw no reason why Darwin's and Lamarck's theories should not be brought into closer alliance.

There were <u>obvious reasons for the</u> <u>stronger growth of heterodox thought</u> <u>in the botanical sciences than in the</u> <u>branches of zoology.</u> First, from the very beginning of Darwinian biology, botanists, much more than zoologists, felt uncomfortable with natural selection—which translated into the struggle for existence—as the primary moving force of evolution. Predatory drive as a reality and motor of evolution was much more visible in the animal world than in plant communities. Second, Russia had developed an especially strong tradition in plant physiology as an experimental science. Among biological laboratories, those in plant physiology were the most modern and elaborate; they brought the advanced methods of physics and chemistry to the experimental domain of biological studies. Laboratory research generated myriad questions that did not fall within the framework of Darwinian concerns.

Symbiogenesis became a legitimate and systematic part of biology in the age of rising theoretical uneasiness in evolutionary studies at the beginning of the twentieth century. The awareness of symbiosis as a unique mode of "cooperation" on the interspecific level had a long past but a short history of intensive and methodical study. In Russia, many pioneers in these studies were, not surprisingly, botanists who found Darwinian perspectives narrow and outdated. But not all Russian botanists were in the camp of Darwin's enemies. Fine botanists—for example, K. A. Timiryazev, N. V. Tsinger, and

Foreword

V. I. Taliev—were among the leading Russian Darwinists.

The specific national concern that stimulated research in symbiosis as a general biological theme was the Russian involvement in the study of lichens, an abundant life form and one that covers large stretches of land in the regions of the Siberian tundra and parts of the northern taiga, as well as in many other parts of the world. Lichens occupy a pivotal position in the ecology of the northern reaches of Siberia. Numerous lichen species make the difference in the survival potential of much of Siberia's vegetative cover and many animal species. In winter, reindeer, both wild and domestic, feed exclusively on a species known as reindeer lichen. Lichens became a particularly attractive subject for scientific study after 1879 when A. de Bary, introducing symbiosis as a biological concept, established that they represented an evolutionary product of the symbiotic association of a fungal species and an algal species that normally exist as entirely different taxa. Lichen study constituted only the beginning of investigations of symbiosis as a mechanism of evolution. Other forms of symbiotic association, some newly discovered, came under intensive scrutiny and only gradually became research concerns of a promising branch of biology.

Three Russian scientists deserve credit for establishing symbiogenesis as a full-fledged and legitimate field of inquiry: Famintsyn (1835–1918), an expert in photosynthesis and a distinguished member of the St. Petersburg Academy of Sciences; K. S. Merezhkovsky (1855–1921), professor of botany at Kazan University; and B. M. Kozo-Polyansky (1890–1957), a professor at Voronezh State University.

Although Famintsyn preferred experimental research and observational data and held theoretical ventures to a minimum, Merezhkovsky displayed a strong flair for abstract thought, intuitive insights, and elaborate logical structures, as described in Khakhina's book. He was the first to rely on symbiogenesis, a term he coined, as a basis for major revisions in the classification of the smaller organisms and as a springboard for a major revision of the theory of biological evolution.

Deep involvement in the experimental study of cell membranes, their inner structures and adaptive functions, led Merezhkovsky to formulate an elaborate evolutionary theory: first, that all cells are symbiotic combinations of living units that originally lived independent lives, and second, that the entire living nature is made up of two plasms: mycoplasm (peculiar first to bacteria and then to the more elaborate cyanophyta and fungi) and amoeboplasm, which, emerging later, appeared in the form of minute "noncellular monera." Behind the somewhat antiquated terminological codification of Merezhkovsky's symbiogenetic theory stood a potent idea that invited a novel approach to the evolutionary process in the universe of living nature. Merezhkovsky's idea on symbiogenesis belonged to the future, not to the time in which he lived. As Lynn Margulis and Dorion Sagan put it: "It has taken the biological world over seventy years to catch up with him."[1]

Merezhkovsky accepted the Darwinian view of the growing complexity of internal structures and physiological processes as the most reliable index of evolutionary "progress," but he did not accept Darwin's thoughts on natural selection as the main factor of evolution. He preferred the cooperation of symbiotic associations to the "conflict" of Darwin's natural selection. The Russian community of biologists was involved in a major confrontation between the champions of new genetics and classical Darwinism. Despite fundamental differences in their general views, it found little use for Merezhkovsky's symbiosis as a mechanism of evolution.

In his attitude toward Darwin's evolutionary legacy, Kozo-Polyansky differed from Famintsyn and Merezhkovsky.[2] Although Famintsyn professed that his theory complemented Darwin's ideas and Merezhkovsky acknowledged that Darwin's and his theories were basically incompatible, Kozo-Polyansky was particularly anxious to present his theory of symbiogenesis as an integral part of the theory of evolution as presented by Darwin. He identified himself as a consistent and thorough Darwinist and waged a relentless war against all kinds of anti-Darwinian stirrings in biology. The acceptance of the ideas of symbiosis and symbiogenesis depended on the acceptance of the concepts built into natural selection and the struggle for existence. He was convinced that the reality of evolution rested to a large extent on symbiotic associations selected for in the struggle for existence. A corresponding member of the Academy of Sciences of the USSR, V. L. Ryzhkov was ready to cite concrete examples to illustrate the work of symbiosis as a factor of evolution.[3]

Kozo-Polyansky's effort to merge the idea of new biology with dialectical materialism attracted little attention, even in Marxist circles. Perhaps Marxist scholars found it much easier to drop the issue of symbiogenesis altogether than to become involved in the arduous task of reconciling the species cooperation emphasized by Kozo-Polyansky with the species conflict built into Darwin's theory. A typical biologist of this time was inclined not to tamper with Darwin's theory. During the 1920s, Russian biologists played a vital role in the development of population genetics in an effort to effect a synthesis of mutation theory with Darwin's principle of natural selection.

1. Margulis, L., and Sagan, D. 1986. *Microcosmos.* Summit Books, New York, 119.

2. Khakhina, L. N. 1983. Problema simbiogeneza: Osnovnye etapy razrabotki problemy. In *Razvitie evoliutsionnoi teorii v SSSR (1917–70-e gody),* ed. Mikulinsky, S. R., and Polyansky, Iu.I. Leningrad, 426.

3. Ryzhkov, V. L. 1966. Vnutrikletochnyi simbioz. *Priroda* 3:9–17.

Foreword

In *A New Principle of Biology* (1924), Kozo-Polyansky presented examples of the wide occurrence of symbiosis in the world of living nature. He made it clear, however, that he did not consider symbiogenesis to be the exclusive factor of evolution. It was obvious that he wanted to recognize the power of the struggle for existence beyond the limits of symbiosis.

The studies of Famintsyn, Merezhkovsky, and Kozo-Polyansky laid the foundation for a burgeoning interest in symbiosis, which in the following decades helped to illuminate the key biological problems related to the inner dynamics of species and to the mechanisms of evolution and which employed many methods, including those of physiology, cell biology, and molecular biology. Despite the originality of their theoretical constructions and the breadth of their knowledge and experimental skills, the three scientists fared poorly. They received much more national recognition for the contributions they made to other fields in biology—Famintsyn for the study of photosynthesis, Merezhkovsky for the systematics and morphology of "lower" plants, and Kozo-Polyansky for the defense of classical Darwinism and the application of Haeckel's biogenetic law to the world of plants. Their ideas were exceedingly slow in reaching the West, where the interest in symbiotic processes occupied a somewhat broader but more diffuse front.

In the Soviet Union, the recognition of the scientific merit of the theory of

symbiogenesis was slow and painful. A quick way to identify attitudes of Russian biologists is to compare entries on the subject in the three editions of the *Great Soviet Encyclopedia*. The 1945 volume of the first edition in which symbiogenesis is an entry refers to the contributions of Famintsyn, Merezhkovsky, and Kozo-Polyansky but expresses a thoroughly negative attitude toward symbiogenesis as a mechanism of evolution. The author of the article, L. Kursanov, states that the theory of symbiogenesis was not confirmed by a single experiment and could not be recommended even as a "working hypothesis."[4] The second edition—most volumes of which were published in the heyday of Lysenkoism (the 1950s)—solved the problem of symbiogenesis by omitting it from even the general article on symbiosis. The third edition of the encyclopedia (in the 1970s), signaled a strong reversal: It carried an article on symbiogenesis, presenting it as the theory that "many eukaryotic cell structures . . . originate as a result of the prolonged symbiosis between eukaryotes and prokaryotes, for example, bacteria and blue-green algae."[5]

The article considered Merezhkovsky, Famintsyn, and Kozo-Polyansky to be the founders of the theory of symbiogenesis, and named A. L. Takhtad-

4. Kursanov, L. 1945. Simbiogenez. *Bol'shaia sovetskaia entsiklopediia*, 51:126.

5. *Great Soviet Encyclopedia*, 3d ed. 1979, s.v. "symbiogenesis."

zhyan, a leading Russian botanist, Margulis (USA) and J. D. Bernal (Great Britain) as its most eminent modern champions. Takhtadzhyan helped to improve the status of symbiogenetic studies in the Soviet Union by putting his great reputation and authority behind it. In "Four Kingdoms of the Living World," published in *Priroda* [Nature] in 1973, he analyzed Western and Russian studies, which led him to conclude that "it has become perfectly obvious that the symbiotic theory of the origin of the eukaryotic cell, previously received with suspicion or rejected, has acquired many supporters in recent years."[6] Modern science, particularly molecular biology, had rejected some of the classical explanations of the origin of eukaryotic cells but had generally accepted classical views on "the kinetic center, chloroplast and mitochondria."[7]

The place of the theory of symbiogenesis in modern biological thought extended even farther. Liya Khakhina (1973) published the first systematic historical survey of the development of this theory in czarist Russia and the Soviet Union, indicating not only its chief architects but also its critics.[8] The monumental *Development of Evolutionary Theory in the USSR,* published in 1983, was the first book-length survey of the history of a leading branch of

Soviet biology to include an entry on symbiogenesis, which states that modern advances have led to the recognition of symbiosis as a "probable factor" in the evolution of eukaryotes.[9] A modern college textbook on Darwinism states that there was a possibility that the emergence of the first animals with a distinct cell nucleus and mitochondria was connected to symbiotic activities.[10] The *Biological Encyclopedic Dictionary* (1986) is both positive and cautious. It acknowledges both the rising importance of the theory of symbiogenesis and its continued controversial nature.[11]

Two books have played an important role in helping to consolidate and strengthen the interest of Soviet biologists in the theory of symbiogenesis: Khakhina's *Concepts of Symbiogenesis* (1979), presented here in an English translation, and Margulis's *Symbiosis in Cell Evolution* (1981), published in a Russian translation in 1983.[12] While Khakhina's book presents a historical analysis of the gradual maturation of the idea of symbiosis as a factor of evolution, Margulis's volume provides a systematic analysis of the full spectrum of scientific arguments in favor of accepting symbiogenesis not as a

6. Takhtadzhyan, A. L. 1973. Chetyre tsarstva organicheskogo mira. *Piroda* 2:27.

7. Ibid., 23.

8. Khakhina, L. N. 1973. K istorii ucheniia o simbiogeneze. *Iz istorii biologii* 4:63–75.

9. Khakhina, L. N. Problema simbiogeneza. op. cit. 421–35.

10. Paramonov, A. A. 1978. *Darvinizm.* Moscow, 148–49.

11. 1986. Symbiogenez. *Biologichesky entsiklopedichesky slovar.* Moscow, 574.

12. Margulis, L. 1983. *Rol' simbioza v evoliutsii kletki.* Moscow.

Foreword

mere probability but as an established and fertile fact of modern scientific thought.

Alexander Vucinich
Professor of History and Sociology
of Science, Emeritus, University
of Pennsylvania, Philadelphia

EDITORS' INTRODUCTION TO THE ENGLISH TEXT

By now nearly every scientist involved in the life sciences acknowledges that eukaryotic cells (those of plants, animals, fungi, and protoctists) evolved from the symbiotic union of two or more types of once free-living prokaryotic microbes (bacteria). This simple idea with far-reaching implications languished for decades in the West, entirely apart from mainstream scientific thought.

The parallel between the prolonged rejection by important biologists of the idea of a symbiotic origin of the eukaryotic cell and the similar rejection by mainstream geologists of the idea of continental drift (revived as a consequence of plate tectonics) is notable. In the case of continental drift, the simple idea was the apparent fit between the adjacent coastlines of South America and Africa. The general assessment of both the idea of the symbiotic origin of the eukaryotic cell and continental drift began to change drastically at nearly the same time, during the late 1960s and early 1970s.

Why did these two paradigm shifts—both the geological and biological—occur so nearly simultaneously? Do these parallel developments have to do primarily with the historical-social and technological context of post–World War Europe and North America, in the same way that Darwin's work has been linked to the political, social, religious, and industrial history of Victorian Britain (Bynum 1991), or did Thomas Kuhn's "Structure of Scientific Revolutions" (1962) have a catalytic effect on both of these twentieth-century scientific revolutions, which followed only a few years after its publication?

At this moment, the Russian people are struggling with their second political revolution of the century. Liya Khakhina's book is a fascinating review of a Russian scientific breakthrough from an earlier age of political revolution. For reasons of language and isolation, the new scientific paradigm fabricated during the first Russian revolution failed to gain a wide audience in either the East or the West until new techniques in biology provided the tools to test Lynn Margulis's assertions that nucleated cells are complex unions of prokaryotes (Sagan 1967; Margulis 1970, 1981).

The illuminating concept of symbiogenesis compares with plate tectonics in making sense of a bewildering quantity of otherwise confusing facts. A book detailing the acceptance of the idea of the symbiotic origin of the eukaryotic cell and Margulis's role in this scientific revolution needs to be written. In lieu of that, the early days of her symbiosis research program are recalled below. The following pages also review the contemporary status of *serial endosymbiosis theory* (SET) detailed in the two articles we wrote together (1990a, 1990b).

Impressed with the frequency of cases of non-Mendelian heredity in a world of nuclear inheritance (for example, photosynthesis mutants of plants and algae, "petites" in yeast, cortical inheritance in *Paramecium,* and so on), as a genetics student at the universities of Wisconsin (1957–60) and California, Berkeley (1960–65), Lynn Margulis explored the early literature for clues to explain cytoplasmic heredity. That there were no "naked genes" outside of cells (or even in them) became clear, and evidence for the presence of bacterial genetic systems inside eukaryotic ones became equally obvious. Based on studies of cytoplasmic heredity and consequent predictions of organellar DNA, she wrote (under the name Lynn Sagan) the statement of the origin of nucleated (*mitosing*) cells from bacterial symbiotic associations in the year 1964–65. James Danielli, co-originator of the lipoprotein-bilayer-membrane theory, published Margulis's work in the *Journal of Theoretical Biology* after approximately fifteen rejections of various transformations of the manuscript (Sagan 1967).

The expanded version of the theory of the origin of eukaryotic-cell organelles (mitochondria, plastids, and centrioles in nucleocytoplasm) was developed into a book originally under contract, and then rejected, by Academic Press. Eventually it was published by Yale University Press (Margulis 1970). The theory itself was supported scientifically (never financially) during the 1970s and 1980s by many new results from molecular biological, genetic, and ultrastructural studies. The revised version of the work became the monograph *Symbiosis in Cell Evolution* (Margulis 1981, 1992).

The theory of the evolution of nucleated cells from a series of bacterial symbioses was named serial endosymbiosis theory by F. J. R. (Max) Taylor (1974). A Canadian marine biologist and expert on dinomastigotes, Taylor attempted, with limited success, to develop the details of the contrasting nonsymbiotic origin of eukaryotes. As an intellectual exercise, he described the origin of organelles by "direct filiation" or autogenesis (Taylor 1976). He also distinguished different versions of cell symbiosis theory: the xenogenous view of organelle origin, that is, the serial endosymbiosis idea.

Taylor collected data on the possible origins of mitochondria (from respiring bacteria) and plastids (from cyanobacteria). The concept of the origin of plastids (but not mitochondria) or the origin of both plastids and mitochondria by symbiosis was labeled the "mild" version of the SET. Taylor called

Margulis's theory the "extreme" version because she had insisted that the cilia kinetosomes and related microtubule organelles (undulipodia and their generative structures) were, like mitochondria and plastids, also of symbiotic origin from motile bacteria. As far as Margulis knew at the time, hers was an entirely original contribution. In the 1960s, cell-symbiosis ideas were still disreputable. Some scientists, like Hans Ris (Margulis's professor at the University of Wisconsin, Madison), knew well E. B. Wilson's 1928 masterpiece, *The Cell in Development and Heredity,* and saw to it that articles in the early 1970s on the discovery of DNA outside the cell nucleus mentioned earlier ideas of organellar origin by symbiosis. Although some authors cited historical views of cell symbiosis and presented ideas of the symbiotic origin of mitochondria and plastids originally brought forth by American (for example, Wallin 1927) and French (such as Portier 1918) authors, these extremely skeptical scientists of the 1960s and 1970s tended to discount hereditary endosymbiosis as a mechanism of evolution (Margulis 1992, chap. 3).

Indeed, Max Taylor and Lynn Margulis were conscious of both prejudice against and ignorance of cell symbiosis theories by most experimental biologists. Yet neither of them had knowledge of our eastern European predecessors: the "symbiogeneticists," such as the Russians Konstantin Sergeevich Merezhkovsky (1855–

1921) and Boris M. Kozo-Polyansky (1890–1957), who from the late nineteenth century until the 1950s developed detailed proposals for the origin of evolutionary novelty, including cell organelles, by hereditary symbioses. Although she was superficially acquainted with some of the names of early Russian and German biologists, her knowledge of the history of the ideas she was espousing was woefully limited.

In a fellowship application to the Eastern European–Soviet Studies Program of the National Academy of Sciences (USA) (1987–88), Margulis was finally in a position to propose research into the origins of the obscure concept of symbiogenesis in pre–Soviet Russia. She planned the fellowship to coincide with the residence of her daughter, Jennifer Margulis, at Leningrad University. She fully expected to have to visit the stacks at Kazan University in search of the intellectual history of Merezhkovsky and his correspondence with the West. Because she neither reads nor speaks Russian, she brought Jennifer as translator so that she would be able to use the time to develop a plan to recover the original papers of Merezhkovsky, Kozo-Polyansky, and the other symbiogeneticists, such as Andrei Sergeevich Famintsyn (1835–1918).

During the visit, in April 1989, she lectured at Moscow State University and mentioned her interest in the concept of symbiogenesis in evolution. A talkative, busy crowd of the curious surrounded her afterward, enduring the usual problems of language incomprehensibility. She felt a pressure on her hand; a small paper-wrapped package was placed in it warmly but firmly by a young earnest woman with whom she had absolutely no words in common. She remains, alas, anonymous. On returning to the Hotel Russiya, to her delight she found in the package two slim volumes: the Russian version of the one you have in your hand now, *Concepts of Symbiogenesis* by Liya Nikolaevna Khakhina, and *History of Concepts in Evolutionary Theory* (1975), edited by K. M. Zavadsky, chapter 1 of which by Khakhina discusses Merezhkovsky's theory of symbiogenesis. To her surprise and relief, she found that Khakhina, employed by the History of Science section of the Academy of the USSR since the death of her colleague Zavadsky (1910–77), had been studying in a professional way what she planned to skim amateurishly. The second book, containing chapters on neo-Darwinism and developmental biology, has yet to be translated.

With luck, Jennifer was able to locate Khakhina's telephone number in Leningrad and make a date to see her. The three women spent a blessed hour in the lobby of the Pribaltiskaya Hotel, some two hundred meters from the seven-story building in which Jennifer was housed in two small rooms with five other Soviet women students. From this frantic conversation about "big ideas in biology," came the decision to translate this book.

Khakhina's main point was fascinating. The early symbiogeneticists believed Darwinism and natural selection to be useless or irrelevant to the concept of evolutionary change. Mainly from botanical and marine biological studies, they deduced that increasingly intimate relations among symbiotic partners led to symbiogenesis: a living together in physical association of organisms of different species wherein the partner organisms become fully integrated. The ability to make these deductions was more or less a direct outcome of the Russian scientific reaction to the Darwinian revolution—particularly, as described by Alexander Vucinich (1988), the strong influence of Lamarckian thought and the rejection of social Darwinism.

For these early biologists, symbiogenesis was the main source of evolutionary novelty. The work of Famintsyn and Merezhkovsky, of the German scientist Richard Altmann, who spoke of theoretical "bioblasts" as elements of the cell, and of A. F. W. Schimper and A. Meyer, discoverers of the plastids of plant cells, was reviewed in a historical context by E. B. Wilson (1928). By introducing and then criticizing the concepts of symbiosis in evolution in the context of the bacterial origins of mitochondria and plastids, Wilson ensured that these concepts were at least mentioned in the West. The much younger associate of Famintsyn and Merezhkovsky, Kozo-Polyansky, however, was not known to Wilson or to anyone else in the English-

Editors' Introduction

speaking world interested in the role of symbiosis in evolution. Yet, according to Khakhina, it was Kozo-Polyansky who brought together the two essential lines of thought: he realized that symbiosis was the source of evolutionary novelty but that natural selection, in the ordinary Darwinian sense of the failure of any population to reach its biotic potential, acted on emerging and tightening symbiotic associations. In Khakhina's opinion, it was Kozo-Polyansky who developed the modern view of symbiogenesis, even though he received almost no credit for it, since his book *New Concepts of Organisms* (1924), was unknown outside of the Russian-literate world. This fact was made known to Margulis at the 1975 International Botanical Congress by the superb botanist Armen Takhtadzhyan, who told her that Kozo-Polyansky, not Margulis, first suggested the concept of the symbiotic origin of undulipodial motility! As you will see, Takhtadzhyan was correct.

Upon returning home Margulis contacted her colleague, geologist Mark McMenamin, about the Khakhina history book, since she knew of his familiarity with Russian and his interest in evolutionary discontinuities as recorded in the fossil record. Having just published a book with his wife, Dianna McMenamin, on the emergence of animals in the fossil record (McMenamin and McMenamin 1990), he was acutely aware of Russian and other relevant scientific literature that Americans and other English-only readers

tend to ignore. With the help of Russian language expert Stephanie Merkel (Cornell University) and émigrée student of biology from the Soviet Union Maria Shteynberg (Hartford, Connecticut), McMenamin assisted in the translation and adaptation of the work for an English-speaking audience. We view this book as an essential aspect of the continuing international dialogue that attempts to evaluate the importance of symbiogenesis as an evolutionary mechanism (Margulis and Fester 1991).

Symbiogenesis, the evolution of novelty by the integration of partners belonging to different taxa—resulting from protracted physical association—had been a principle of evolution (at least in Russia) since the late nineteenth century. The term was coined by Merezhkovsky in part to explain the presence of very similar greenish photosynthetic units in heterotrophs as diverse as hydra, paramecia, and plant cells. Merezhkovsky, with his theory of two plasms, and his senior colleague from St. Petersburg, Famintsyn, were the most successful articulators of the theory of chloroplast origins as a specific example of the general symbiogenesis principle.

These keen Russian biologists suggested, in contemporary terms, that plastids originated as captive cyanobacterial symbionts in heterotrophic cells. Although dialogue, mutual criticism, and disagreement existed between these professional biologists, both rejected natural selection, and both thought symbiosis to be crucial as an evolutionary mechanism (Khakhina 1979). It is Kozo-Polyansky, however, who, in F. J. Dyson's (1985) terms, was our most "illustrious predecessor" in claiming the symbiotic origin of centrioles. Kozo-Polyansky's position on eukaryotic cell motility was clear.

Bodies known as blepharoplasts, anchored in the plasma of the cell and bearing a flagellum [undulipodium] or several flagella, come out from the cell interior. Blepharoplasts occur in mastigotes, flagellated cells of sponges, and also spermatozoa. Apparently they occur in plants as well as in animals.

It is recognized widely that blepharoplasts [kinetosomes] are a type of centrosomes (or centrioles); the first may be transformed into the second and vice versa as seen microscopically in live cells. Furthermore, the divisions of cells are synchronized . . . with the divisions of blepharoplasts (in the role of centrosomes or centrioles)—that is, the motile flagella (or flagella-producing, or flagellated organelles or partners of the cell). . . . Without question, at least the suspicion of the bacterial nature of these kinetoplasmatic [motility] organelles . . . is fully warranted. (Kozo-Polyansky 1924, 56–57)

Furthermore, in his book *New Concepts of Biology*—unlike his own illustrious predecessors Merezhkovsky and Famintsyn—Kozo-Polyansky embraced fully the natural selection concept of Charles Darwin. He recognized that natural selection plays a crucial part in symbiont integration: if symbiosis is the author, natural selection

is still the editor. Therefore, Kozo-Polyansky described the power of symbiogenesis in evolution; he used the term with the original meaning of its inventor.

In a definitive passage, Khakhina (1979) cites Merezhkovsky's evaluation of the term symbiogenesis:

The role of symbiosis in evolution, an idea he developed over the course of seventeen years (1903–1920), stands out in all of Merezhkovsky's work. Merezhkovsky suggested the term "symbiogenesis" in 1909 and later gave a detailed definition: "I called this process symbiogenesis, which means: the origin of organisms through combination and unification of two or many beings, entering into symbiosis" (Merezhkovsky 1920, 65). Symbiogenesis as an evolutionary principle permitted one to pose and then solve the question of cell origins and organism evolution. Merezhkovsky formulated the evolutionary concept that he called "the theory of symbiogenesis." "So many new facts arose from cytology, biochemistry, and physiology, especially of lower organisms, that an attempt once again to raise the curtain on the mysterious origin of organisms appears desirable. I have decided to undertake such an attempt, and my present work . . . consists in a preliminary exposition of a new theory on the origin of organisms, which, in view of the fact that the phenomenon of symbiosis plays a leading role in it, I propose to name the theory of symbiogenesis" (Merezhkovsky 1909a, 7–8).

But the language barrier proved insurmountable: although the works of both Merezhkovsky and Famintsyn were cited by German and English speakers (Wilson 1928), Kozo-Polyansky's contribution is still vir-

Editors' Introduction

tually unknown outside the Russian Federation. The American (University of Colorado, Denver Medical School) anatomist Ivan E. Wallin (1883–1969) developed the term *symbionticism,* an extremely similar theory of the role of microbial symbiosis in speciation and cell organelle origin, in the total absence of interaction with his contemporary Russian symbiogeneticists. Wallin was heartily disdained; his last article in 1965 was rejected by the journal *Science* (see the Appendix).

Although Paul Portier in France (1918), Umberto Pierantoni in Italy (1948), and Paul Buchner in Germany (1965) were all sympathetic, in varying degrees, to symbiosis as a mechanism for the generation of evolutionary novelty, these authors did not hold major positions in the world of science. The Russians, however, were leaders: Merezhkovsky was a professor at Kazan University, the second most important university in Russia (after Moscow), and Famintsyn was founder of the St. Petersburg Plant Physiology Laboratory. Nevertheless, symbiogenesis was virtually unknown outside of the Russian-speaking world where the concept, if not ignored, was either labeled controversial or ridiculed (for example, Lumière 1919). Wallin's term *symbionticism* was used only by him, the first and last symbionticist.

If the extreme version of the serial endosymbiosis theory of cell origins is correct, then the inescapable inference is that we are all symbiogeneticists. All animals, fungi, and protoctists had at

least three kinds of microbial ancestors, and all plants and algae had at least four. Evolutionary innovation for all eukaryotes involved far more than the accumulation of mutations: it required the integration of heterologous genomes. If all animal cells have at least three ancestors and all plant cells at least four, how many heterologous ancestors has a human, a cow, or a weeping willow? Not only will the concept of *individual* be replaced with that of symbiotic complex for all animals, but since all eukaryotes harbor heterologous DNA's from various sources, both the sciences of eukaryotic evolution and of developmental biology transform. They become special cases of applied microbial community ecology.

Symbiogenesis as an evolutionary principle is under reconsideration (Maynard Smith 1988; Law 1989; Smith and Douglas 1987; Margulis and Fester 1991). Of course, that new organisms evolve by symbiont integration is not entirely new to the English-speaking world. It is the explanation of choice for the origin of lichen structure, termite wood-digestion, and the luminous organs in leognathid fishes. But no longer will statements such as "all lichens are symbioses between algae and fungi" be limited to lichens. Future cell-biology texts must begin with a description of how all eukaryotic cells are communities derived from coevolved symbiotic bacteria. Future general biology textbooks, in addition to explaining why *protozoa* (as "single-celled animals") is no longer a valid taxon, must laud the enormous success of the chloroplasts (phototrophic, oxygenic bacterial endosymbionts) that tint our planet green.

The recent spectacular illustrations of kinetosomal-centriolar DNA reported from the laboratory of David Luck at Rockefeller University (Hall, Ramanis, and Luck 1989) have been attacked vigorously by two groups of investigators (Johnson and Rosenbaum 1990, 1991; Kuroiwa et al. 1990). If Luck and his collaborators are shown to be correct, their compelling experimental evidence will establish definitively all original postulates of the extreme version of the serial endosymbiosis theory. The symbiotic origin even of undulipodia will be accepted as fact by all biologists of all countries. Even now, independent of whether or not undulipodia originated as symbionts, the concept of symbiogenesis must be incorporated into modern evolutionary theory in much the same way that plate tectonics has become the central organizing concept for contemporary geology. Biology, the science of life, needs therefore what might be referred to facetiously as a *postmodernist synthesis,* which acknowledges the discontinuous nature of evolution by symbiogenesis. Such a synthesis will challenge the standard neo-Darwinian view so popular among English-speaking evolutionists of today.

Neo-Darwinian thinking emerged from the fusion of two intellectual lineages: (1) the patterns of the "vertical" transmission of genes (from parent to offspring), primarily in diploid, obligately sexual genetic systems typical of animals, and (2) Darwin's ideas of evolutionary change in species through time. The clever application of algebra by G. H. Hardy to the rules for the vertical transmission of diploid "factors" uncovered by Gregor Mendel—performed by R. A. Fisher, J. B. S. Haldane, and S. Wright—led to the "the modern synthesis." The new synthesizers forced a system of constancy (Mendel's factors are *not* changed by transmission) into a system requiring change (of varieties and species through evolutionary time, as argued by Darwin). They set up a rigid framework in which change in the genetic make-up of populations was attributed to a limited set of possibilities: mutation, migration in and out of the population by its members, founder effects of small populations, and the like. That evolutionary change originated from inside the population in question and was necessarily gradual became an unchallenged tenet in biology; "punctuated equilibrium" interpretations of the discontinuities in the fossil record were mostly ignored by population biologists.

Neo-Darwinian thought established a place for everything in evolution. This unstated assumption pervaded, and still pervades, many evolutionary textbooks. The gradualist neo-Darwinian paradigm excludes symbiogenesis except as an oddity of limited interest, mainly to cell biologists and biochemists. Since chemically oriented biologists, biochemists, and cell physiologists tend to dismiss all of evolutionary biology as an untestable historical enterprise unworthy of experimental science, few were concerned directly with the differences in "thought-style" between neo-Darwinians and experimental biologists. Although a limited number of biologists did study the phenomenon of symbiosis with experimental techniques, symbiogenesis as a mechanism to generate fundamentally new features of organisms (such as lichens and bacterial light organs in marine fishes; Margulis and Fester 1991) was ignored or even denigrated by professional evolutionary biologists. As molecular studies have begun to reveal selectively neutral mutations, large genomic rearrangements, vastly different rates of sequence change through time in different proteins and genes, new cytoplasmic genetic systems (some of bacterial origin), molecular drive, and many other phenomena entirely unpredictable by neo-Darwinian preconception, the mood has shifted. A willingness to consider the significance of symbiosis as a general evolutionary mechanism, involved not only in innovation but directly in speciation and morphogenesis, has surfaced recently. Critical of limitations of neo-Darwinian zoocentric dogma and newly sensitive to natural history and its undescribed wonders, we became aware of erudi-

tion on the eastern side of the globe. Studies of evolution are not limited to our own scholarly traditions. In this spirit we have brought Khakhina's historical work into the necessary dialogue about the nature of changing nature.

In this year of the celebration of five hundred years of hemispheric contact, Liya Khakhina's awareness of the rich tradition of views of the importance of symbiosis as an evolutionary mechanism should transcend the Russian language and be available to all interested readers. We, the editors, along with Khakhina, invite you to participate in the continuing East-West geobiological search for an understanding of the evolutionary process. In this spirit of historical inquiry, we bring to light the story of the lone American symbiogeneticist, Ivan Wallin, told by a historian of biology, Donna Mehos.

LYNN MARGULIS
University of Massachusetts, Amherst

MARK McMENAMIN
Mount Holyoke College
South Hadley, Massachusetts
1992

References to
Editors' Introduction

Buchner, P. 1965. *Endosymbiosis of Animals with Plant Microorganisms.* Interscience, New York.

Bynum, W. 1991. Charles: A novel view of the man. *New Scientist* 132 (1792): 54.

Dyson, F. J. 1985. *Origins of Life.* Cambridge University Press, Cambridge.

Hall, J. L., Z. Ramanis, and D. J. L. Luck. 1989. Basal body/centriolar DNA: Molecular genetic studies in *Chlamydomonas. Cell* 59:121–32.

Johnson, K. A., and J. L. Rosenbaum. 1990. The basal bodies of *Chlamydomonas reinhardtii* do not contain immunologically detectable DNA. *Cell* 62:615–19.

———. 1991. Basal bodies and DNA. *Trends in Cell Biology* 1:145–49.

Khakhina, L. N. 1979. *Concepts of Symbiogenesis.* Akademie NAUK, Leningrad.

Kies, L., and B. P. Kremer. 1990. Phylum Glaucocystophyta. In *Handbook of Protoctista,* by L. Margulis, J. O. Corliss, M. McMenamin, and D. J. Chapman, eds. Jones and Bartlett. Boston, 152–66.

Kuhn, T. 1962. *The Structure of Scientific Revolutions.* University of Chicago Press, Chicago.

Kuroiwa, T., T. Yorihuzi, N. Yabe, T. Ohta, and H. Uchida. 1990. Absence of DNA in the basal body of *Chlamydomonas reinhardtii* by fluorimetry using a video-intensified photon-counting system. *Protoplasma* 158:155–64.

Law, R. 1989. New phenotypes from symbiosis. *Trends in Ecology and Evolution* 4:334–35.

Lumière, A. 1919. *Le Myth des Symbiotes.* Masson, Paris.

Margulis, L. 1970. *Origin of Eukaryotic Cells.* Yale University Press, New Haven.

———. 1981. *Symbiosis in Cell Evolution: Life and Its Environment on the Early Earth.* W. H. Freeman, San Francisco.

———. 1992. *Symbiosis in Cell Evolution: Microbial Communities in the Archean and Proterozoic Eons.* 2d ed. W. H. Freeman, New York.

Margulis, L., and R. Fester, eds. 1991. *Symbiosis as a Source of Evolutionary Innovation: Speciation and Morphogenesis.* MIT Press, Cambridge.

Margulis, L., and M. McMenamin. 1990a. Marriage of convenience: The motility of the modern cell may reflect an ancient symbiotic union. *The Sciences* September/October:31–37.

———. 1990b. Kinetosome-centriolar DNA: Significance for endosymbiosis theory. *Treballs de la Societat Catalana de Biologia* 41:1–9.

Maynard Smith, J. 1988. Evolutionary progress and levels of selection. In *Evolutionary Progress,* ed. M. H. Nitecki. University of Chicago Press, Chicago.

McMenamin, M. A. S., and D. L. S. McMenamin. 1990. *The Emergence of Animals: The Cambrian Breakthrough.* Columbia University Press, New York.

Merezhkovsky, K. S. 1909. *The Theory of Two Plasms as the Foundation of Symbiogenesis, New Knowledge on the Origins of Organisms.* Proceedings of Studies of the Imperial Kazan University. Publishing Office of the Imperial University. (In Russian.)

———. 1920. La plante considerée comme un complexe symbiotique. *Societé des Sciences Naturelles de l'Ouest de la France, Nantes, Bulletin* 6:17–98.

Pierantoni, U. 1948. *Trattato di Biologia e*

Zoologia Generale. Humus, Naples.

Portier, P. 1918. *Les Symbiotes*. Masson, Paris.

Sagan, L. 1967. On the origin of mitosing cells. *Journal of Theoretical Biology* 14:225–74.

Smith, D. C., and A. E. Douglas. 1987. *The Biology of Symbiosis*. Edward Arnold, London.

Taylor, F. J. R. 1974. Implications and extensions of the serial endosymbiosis theory of the origin of eukaryotes. *Taxon* 23:229–58.

———. 1976. Autogenous theories for the origin of eukaryotes. *Taxon* 25:377–90.

Vucinich, A. 1988. *Darwin in Russian Thought*. University of California Press, Berkeley.

Wallin, I. E. 1927. *Symbionticism and the Origin of Species*. Williams and Wilkins, Baltimore.

Wilson, E. B. 1928. *The Cell in Development and Heredity*. 3d ed. Macmillan, New York.

ACKNOWLEDGMENTS

We are grateful to Edward Tripp and Jean Thomson Black for their roles in obtaining permissions and clearing the way for this translation. We thank Robert Coalson, Clifford Desh, Jennifer Margulis, Stephanie Merkel, and Maria Shteynberg for the translation work. We acknowledge the manuscript preparation aid of Eileen Crist, Stephanie Hiebert, Lorraine Olendzenski, and Dorion Sagan, and the financial support of the Richard Lounsbery Foundation, New York, and the College of Natural Sciences and Mathematics of the University of Massachusetts, Amherst, the National Science Foundation, and the U.S. National Academy of Sciences. Without the National Academy Fellowship and the unfettered generosity of the Richard Lounsbery Foundation, this project would not have been possible.

Note on Translation and Transliteration

A nearly literal translation of Khakhina's preface has been sought. For the rest of the book, however, we have used colloquial terms for English-speaking audiences (for example, *community* instead of *biocoenosis* and *microbial community* instead of *microcoenosis*). We have replaced "higher animals" and "higher plants" with the intended taxon (mammals, vertebrates, angiosperms, and so on). We have retained the term *chromatophore* ("colored bodies") to ensure the fidelity of translation; this word sometimes, but not always, refers to chloroplasts. We have also retained the term *arogenesis,* which may be unfamiliar to English-language readers. *Arogenesis* refers to the evolution of complexity in the organic world throughout life's history. The most similar word in the West is *complexification,* as employed by Teilhard de Chardin. Since *arogenesis* lacks the teleological connotations intended by Teilhard, however, we believe that it is important to retain Khakhina's term.

Complexification and arogenesis are considered to be synonymous throughout the text. Both terms are used by the editors without teleological implications; no evolutionary advancement or improvement can necessarily be inferred merely from their use. We quote, however, Dr. Khakhina's response on reading this comment on the translation in the manuscript:

> Arogenesis refers to the trend of evolution proceeding in the direction of a higher level, that is, a more perfect organization (progressive evolution). The principal feature of arogenesis consists of acquisition, accumulation, and perfection of a whole complex of adaptations having great ecological significance. Morphologically and physiologically, arogenesis involves the increase in complexity of the organization of an individual. In non-Russian scientific literature, the counterpart for the term *arogenesis* is the concept *anagenesis* (Bernard Rensch, *Evolution above the Species Level,* Columbia University Press, 1960). The reader is advised to see Rensch for a discussion of the term *anagenesis.*

Thus, although we have tightened most of the prose, using terms in the modern (not the early twentieth century) sense (for example, *cyanobacteria* instead of *cyanophyte, flagellated bacteria* instead of *flagellated cytodes*), of course we strove to keep the text precisely within Khakhina's original meaning and intent. Editors' comments are in square brackets.

We have used system 1 from J. Thomas Shaw's, *Transliteration of Modern Russian for English Language Publications*, Modern Language Association of America, New York (1979), a ten-page pamphlet. Transliteration of all Russian names conforms to Shaw's recommendations.

Interested researchers may contact Mark McMenamin for transliterated titles of particular Russian references cited in Khakhina's book. For those readers interested in sources not fully referenced, please contact the author directly at

The Academy of Sciences of the
Russian Federation
Institute of Natural Science
and Technology
Universitetskaya nab. 5
St. Petersburg, 199034 Russia

Mark McMenamin
Lynn Margulis

L. N. KHAKHINA

Concepts of

Symbiogenesis

A Historical and Critical Study of

the Research of Russian Botanists

P REFACE

During the past few years, the Department of the History and Theory of Evolutionary Studies of the Institute of Natural Science and Technology of the Academy of Sciences of the Russian Federation has been conducting research on the evolution of complexity in the organic world (arogenesis). This research has included studies of the criteria of arogenetic evolution, its basic forms and concepts, the external and internal conditions of its realization, its moving forces, the laws governing it, and so forth.

The problem of the increasing complexity of organization deserves a special place within the problem of progressive evolution. The study of this phenomenon, and particularly of the factors and causes that lead to it, has occupied a significant place in the history of evolutionary biology since Darwin. As one of the methods of increasing the complexity of the organization of individuals, symbiosis is emerging as a probable source of progressive evolution.

In evolutionary theory, the hypothesis that the increasing complexity of organization is realized in the course of historical development by the gradual accumulation of subtle, inherited changes has become well established. The long, cumulative process of such step-by-step changes leads to the differentiation or dissolution of a previously integrated system, the parts of which gradually become more and more independent of one another. The origin of highly complex systems, such as the higher plants and animals, is explained convincingly by the selective accumulation of a number of minute, inherited changes [mutations].

It would, however, have been beneficial for the theory of evolution if it had been proved successfully that a simpler and more effective path to the increasing complexity of individual organization exists. During the last century, the possibility of evolution by the sudden radical "recoinage" of forms, macromutations, or systemic mutations has been discussed, but the reality of such a path not only has not been proved but has become even less probable. More and more data have been accumulated substantiating the severely damaging effects of such mutations, which render their bearers not only badly adapted and noncompetitive but also largely nonviable.

Quite different, however, is the possibility that the complexification of organization occurs through the union of previously prepared blocks in a process of the

1

increasing integration of the components of a symbiotic system—that is, by symbiogenesis. The uniqueness of this method of evolution was precisely described by the academician A. L. Takhtadzhyan (while he was studying the origin of the eukaryotic cell) as a process of the "assembly" of a complex system from largely "prefabricated parts" (1973, 27).

The idea that two or more symbionts could unite gradually to form a single, more complex system that is itself on the level of the organism is very attractive because each of these partners has already demonstrated its fitness over the course of its history. The question that arises is this: How widespread is symbiogenesis as a method of complexification in nature? Is it merely a limited factor of evolution, acting only during the formation of such taxa as lichens, or is it significant as a general factor of evolution, facilitating the origin of entire branches of organisms (for example, eukaryotes)?

This book employs a historical-critical method. It offers an evaluation of the status of the concept in the past and of its development from the position of the modern hypothesis; it takes into consideration new factual material and the basic sources for the study of an entire set of concepts that are essential for the elaboration of any logical system. This system will serve the researcher as a basic reference point. Such analysis has been proved effective, particularly in cases in which the problem has traversed a long and complex path of development, and its further study demands the rethinking not only of its modern status but also of the material and the ideas that have accrued around it. Studies of the problem of symbiogenesis encourage one to think that the application of the method of historical-critical analysis will be useful not only for the summary of work already completed in this field but also for the organization of future research.

As will be shown in this book, the problem of symbiogenesis followed a rather complex path of development. This path was not direct; it was replete with deviations and periods of stagnation. The origin and development of the question of the role of symbiosis in evolution still has not been considered systematically in the literature. I will attempt to fill this gap.

The problem of symbiogenesis is closely connected to Russian evolutionary thought. The idea that symbiosis is a significant factor of evolution was formulated almost simultaneously by A. S. Famintsyn and K. S. Merezhkovsky. These two were the first to attempt to base the theory in logic and fact. The elaboration of this concept from the position of Darwinism is associated with B. M. Kozo-Polyansky. A. A. Elenkin made the most complete critical analysis of symbiogenesis. Further research has been carried out by V. N. Lyubimenko, A. N. Danilov, A. G. Genkel', P. A. Genkel', and V. L. Ryzhkov.

One of the difficulties of this work has been its requirement for materials from quite diverse areas of biology (community ecology, lichenology, the biology of prokaryotes, the physiology of some organelles, molecular biology, and others). Because all of these sources have a direct bearing on the general biological problem of symbiogenesis, I have tried to present them in such a way that they will be understood by biologists of various specialties. All Latin terms are cited exactly as they were given by the original authors.

This book was written under the direction of Kirill Mikhaylovich Zavadsky (1910–77), Doctor of Biological Sciences, professor and honored scientist of the Russian Socialist Federated Republic. His ideas and hypotheses regarding symbiosis as a factor of evolution lie at the heart of this book, which is essentially a record and an elaboration of them. Having lost the opportunity to thank Kirill Mikhaylovich for his many years of directing my research on the problem of evolution, for his enormous help in the writing of this book, and for his uncommon intellectual generosity, the author dedicates this book to his fond memory.

The author also wishes to express her deep gratitude for the valuable remarks and useful advice of Doctor of Biological Sciences M. M. Gollerbach and Doctor of Chemical Sciences S. E. Bresler.

1

SYMBIOGENESIS AS A CONCEPT IN EVOLUTIONARY THEORY

The Status of the Symbiogenesis Concept

The progress of scientific knowledge is related to the advancement of new hypotheses. These hypotheses enable us to explain accumulated facts according to current scientific developments. Hypotheses based on limited material are often insufficiently grounded for the formulation of strictly testable scientific theories. A good working hypothesis demands more evidence and constitutes an important stimulus to seek scientific knowledge: "To the extent that natural science thinks, the hypothesis is the form of its development," wrote F. Engels.*

Scientific hypotheses have various fates. The accumulation of new facts may elevate a hypothesis to the level of scientific theory or lead to the refinement or alteration of the hypothesis. At the same time, the verification of a hypothesis may demonstrate the inadequacy of a previously suggested explanation. Refusing to construct new hypotheses and striving merely to accumulate new facts will always inhibit the process of learning, particularly in general areas of science such as evolution theory.

The development of evolutionary thought has always been accompanied by the formulation of diverse hypotheses. Among them attention should be paid to those that not only adhere to Darwinism but also substantially supplement it. Among these is the hypothesis of symbiogenesis or the idea that the evolutionary process, moving toward ever-more profound relations between organisms in symbiotic relations, can sometimes lead to such a close union that a new formation at the level of the organism arises—a complex form having the attributes of an integrated morphophysiological entity (Khakhina 1972a). Symbiogenesis is therefore itself a basic means of evolution, at the heart of which lies the phenomenon of the ever-increasing integration of the symbionts. Symbiosis acts in evolution as a factor that increases the complexity of organization.

In the classical point of view (Darwin 1939), taxa evolve by the breakup of formerly whole groups into parts, which in turn become isolated from one another and from the original, ancestral group. Another mechanism of evolution,

---

*F. Engels, *The Dialectics of Nature*, in K. Marx, F. Engels, *Collected Works*, vol. 20, p. 255.

however, has been observed—the merging of individuals from two or more different taxa into a single system at the level of the organism. It has been proposed that these two broad methods of evolutionary transformation be called *segregogenesis* and *synthogenesis* (Zavadsky 1967, 1968).

Evolution by synthogenesis includes two basic types of integration: interbreeding (hybridgenesis) and the asexual integration of heterogeneous phenotypes. Proceeding on the basis of the recombination of genetic material in a developed form (that is, in the form of the sexual process), synthogenesis is the most widespread and important [process] for evolution. The significance of the sexual process in evolution has been thoroughly substantiated. Less clear, however, is the evolutionary role of the asexual union of organisms. By this method, individuals of a single species or of different and, as a rule, phylogenetically distant species can be united. In the first case, which may be called homomorphous unification, colonies are formed by a process of reproduction that is not carried through to completion and by the nondivergence of the offspring organisms. With the further development of the morphofunctional integrity of the colony and the parallel weakening of the individuality of its members, individuals of a higher rank may develop, as is the case in the origin of animals (Ivanov 1968). In heteromorphic unification, the second case, symbiotic associations arise in which strengthening interconnections are selected that are beneficial to the individual components and to the integrity of the entire system, which may lead to the formation of a complex individual or organ.

In recent years a number of researchers have supported the idea of the evolutionary complexification of an individual's organization by the union of previously independent and simpler individuals (Zavadsky 1967, 1968; Margulis 1971a; Belozersky and Mednikov 1972; Lybishchev 1972; Takhtadzhyan 1973; Urbanek 1973; Leninger 1974; Fox and Doze 1975; Shaposhnikov 1975; Zavadsky and Kolchinsky 1977; Tchaikovsky 1977; and others).

Analysis of the development of the idea of symbiogenesis in Russian research in botany, as well as the study of its current status, allows us to assert that the idea of symbiogenesis is a legitimate and fruitful working hypothesis in that it stimulates the quest for additional evidence, compelling the reevaluation of previously obtained empirical material and demanding the accumulation of new factual data to verify it. The hypothesis of symbiogenesis stimulates the development of scientific thought. The accumulation of new data may transform this hypothesis into a strictly proven scientific theory with broad biological significance. It is also possible, however, that the need will arise to limit the hypothesis, that its inadequacies will be revealed.

Having considered the existing criticism of the hypothesis of symbiogenesis from this viewpoint, it may be said that prematurely elevating the hypothesis to the status of a completed scientific theory or universal principle of biology or rejecting it as scientifically unworthy are both equally unjust.

The Idea of Symbiogenesis

The hypothesis of symbiogenesis is intimately related to the independent and complex general biological problem of symbiosis. A detailed analysis of the role of symbiosis in evolution is warranted. In this work, however, I review briefly only those aspects of symbiosis that are vital to analysis of the question at hand.

The phenomenon of symbiosis and its causative preconditions have long been the center of attention for biologists, of equal interest to representatives of the most diverse specialties in the field. This study involves researching the forms of the phenomenon and its prevalence in the plant and animal kingdoms, as well as analyzing the nature of the interrelations between the components of symbiotic associations and the adaptations of the symbionts.

A. de Bary (1879) made the first attempt in botany to characterize symbiosis. He defined the phenomenon in a speech at the Congress of German Naturalists and Doctors at Kassel in 1878 (Kamensky 1891). Having typified symbiosis by the relationship between ascomycetes and algae in a lichen, he interpreted the term widely, considering it to include the close, combined life of two heterogeneous organisms, between which may exist various interrelations. Having included within the concept of symbiosis a wide group of natural phenomena, de Bary picked two extreme forms of cohabitation: antagonistic symbiosis—that is, the parasitic form of symbiosis—and mutualistic symbiosis, in which cohabitation is mutually beneficial.

Later, on the basis of de Bary's characterization of symbiosis and the first classification of its forms by P. van Beneden, the two basic approaches to the definition of symbiosis developed. Supporters of the broader approach, who interpreted symbiosis in the most general sense, understood it to be a close, obligate cohabitation of organisms, including all forms of interrelation that affect, either positively or negatively, the survival of the components: commensalism, mutualism, parasitism (van Beneden 1876), helotism (Varming 1901), endoparasitic saprophytism (Elenkin 1902, 1904),[†] and so forth. This point of view was shared by F. Kamensky, A. A. Elenkin, G. Nadson, A. Genkel', A. Danilov, S. Navashin, Z. Katznelson, and many others.

[†]A similar demarcation of the forms of symbiosis was evaluated critically by A. A. Elenkin in 1940. He justly noted that all of the categories mentioned can be observed within the limits of one or another symbiosis (for example, lichenous), and therefore such distinct classifications are essentially formal and contribute little to an explanation of the substance of symbiotic relations.

The more narrow approach to the definition of symbiosis was based on the principle that members of the association benefit each other. Symbiosis was reduced to a representation of cohabitation in which the components benefit somehow from the union: for example, the mutual supplement of metabolic products, mutual defense, improved conditions of respiration or of movement, and so forth. This is how the term was defined in the works of N. M. Shemakhanova (1962), V. N. Sukachev and N. V. Dylis (1964), E. Odum (1968), C. D. Darlington (1972), among others, as well as in the following dictionaries: *Webster's International Dictionary* (1946), *A Dictionary of Scientific Terms* (1960), and *A Dictionary of Biology* (1962). This understanding of symbiosis excluded all cases of parasitic relations.

The attempt by Soviet animal parasitologists to interpret the biological essence of symbiosis in connection with the analysis of the nature of parasitism (Pavlovsky 1934; Afanas'ev 1937; Filipchenko 1937; Moshkovsky 1946; Dogel' 1962) was an important milestone in elaborating the problem of symbiosis. Parasitism was characterized as an association in which the host is not simply a cohabitant but forms the environment that the parasite inhabits and to which it is adapted: that is, essentially, the host regulates the relations between the parasite and the surrounding external environment. Symbiosis is defined as a type of association in which both members "jointly take part in the regulation of relations with the external environment" (Dogel' 1962, 13–14). In symbiosis, each partner supplements its own deficiencies in features necessary for adaptation to particular conditions by uniting with another organism possessing these features. Symbiosis, therefore, was understood as a special means of adapting an organism to the external environment for mobility, defense against enemies or harmful factors of the environment, and so forth, which is achieved through the unification of organisms. With this approach, symbionts are contrasted with free-living organisms, not parasites.

Finally, the concept of symbiogenesis is variously interpreted. The term is used to represent all community relations among populations of different species within ecosystems, including both the positive (see, for example: Odum 1968; Odum 1975; Dazho 1975; and so forth) and negative relationships of organisms (for example, antagonistic nutrient disjunctive symbiosis, MacDougall 1935).

In addition to these definitions, symbiosis is characterized as an association distinguished by great stability and by the profound association of its components. This representation, formed in the last quarter of the nineteenth century, supposes that integrated, associated symbionts form live systems with a structural and functional identity comparable to any living organism. J. Reinke (1873) was one of the first to express this concept of "consortium," which was developed from the example of the lichen symbiosis (Reinke 1894, 1895). The association of two organisms, Reinke asserted, in this case forms such a unity, a consortium, in which the components merge into a single, indivisible whole that functionally and morphologically corresponds to the features of any green plant.

Many biologists held a similar view of symbiosis (Famintsyn, see chapter 2; Varming 1901; Merezhkovsky, see chapter 3; Bernard 1909; Lyubimenko 1916, 1923; Kozo-Polyansky, see chapter 5; Ryzhkov 1966; and others).

It seems that symbiosis is related to a category of relations that are formed between representatives of different phylogenetic groups. Symbiosis is a form of interaction between heterogeneous organisms in which joint existence is beneficial for the individuals and secures for the partners an essential selective advantage. The value of symbiosis is defined by the fact that, upon entering into an association, an organism becomes better adapted to the environment because of the use it makes of peculiarities already possessed by its partner. In some cases, symbiosis is significant for improving adaptation for feeding; in others for securing high rates of respiration, intensifying metabolic processes, movement, defense from enemies, and so forth.

Given this understanding, the benefit realized from the association by one or both partners is naturally included in the definition of symbiosis and does not require special stipulation.

The only certainty is that there are many diverse forms of symbiosis. Moreover, all cases of symbiotic relations are characterized by a single, common feature—the presence of a definite (sometimes stronger, sometimes weaker) association between the members.

The Transition from Ecological to Physiological Relations

At the base of the hypothesis of symbiogenesis lies the possibility of the gradual transformation, through evolution, of two or more individuals that are bound by community relations into a system that functions as single unit. A new system on the level of the organism may arise from the phenomenon of the community interaction of organisms that do not exchange genetic information but between which exists a close biological and evolutionary relationship. The evolutionary transformation of organisms may proceed by the transition of the components of a symbiotic coevolutionary system into parts of a single organism that interact (co-adaptation).

The progressive connection and integration of the components of a symbiotic association do not always result in unification or the origin of a single organismic system. In most cases, the hurdles to overcome on this path of development are so great that, despite protracted—in the historical sense—coexistence, movement in the direction of a new complex organism does not occur.

Comparing forms of association by the extent of the relationship between the partners reveals a series of distinguishable stages in the union of organisms.[5] At the beginning of this series stand those cases of association that are characterized by a relatively small degree of unity of the partners.

Such, for example, are the community interrelations between ants and aphids, which are distinguished by great diversity. "Not all ants visit aphids," noted the prominent researcher of this type of symbiosis, A. K. Mordvilko, "but among those that do, some use aphids as only one of several sources of nourishment, while others use aphids as their primary, or even as their only, source of nourishment. Further, there are various types of aphids: some are zealously visited by ants, some rarely, and others not at all" (1936, 46). The association of ants and aphids is beneficial for both. Each has worked out special adaptations for this relationship. Thus, when an ant visits an aphid that lives on the aboveground parts of plants, it rapidly strikes the aphid with its antennae. The aphid in response releases a drop of excrement, which the ant licks off. Aphids, when visited by ants, do not spurt out these drops but hold them at the end of the abdomen on the perianal hairs that are located primarily on the anal sternite and, to a lesser extent, on the tergite. The aphids benefit from these visits in that their colony is kept almost completely free of enemies, which are destroyed by the ants. Even closer relationships have been observed between root aphids and underground ants. Equally distinctive are the relations between leaf-cutter ants and the fodder fungi *Septobasidium* (Batra and Batra 1967), those between flowering plants and pollinating insects, cases of the so-called cleaning symbiosis of fish (Limbaugh 1961), and other similar associations formed between members of a single consortium. The weakest connection between members of an association, as V. A. Dogel' (1962) noted, is observed in some cases of consortia. All unions of organisms that are similar to the examples above are cases of typical community relations.

The next stage of the unification of organisms includes association with an extensive relationship between its components. Such symbioses are more closely and strongly bound associations. The symbiosis between the hermit crab (*Pagurus arroser*) and the sea anemone (*Calliactus parasitica*) or among several species of epiphytes, as well as the extracellular symbiosis of the colonial cyanobacteria *Woronichinia naegeliana*, the mucus of which is inhabited by other

---

[5]This idea had already been precisely formulated by A. S. Famintsyn in 1889. All forms of symbiosis, he noted, are transitional stages between two extreme forms, which are clearly distinguished from one another by the extent of the connection between the partners. "In many cases, the connection between symbiotic organisms is very loose, and the independence of each partner is so obvious that it is recognized. . . . In other cases, special laborious experiments are required to prove that the symbiotically originated form breaks down into its constituent parts and to show their independence" (Famintsyn 1889a, 1).

cyanobacteria—*Lyngbya endophytica* or *Synechocystis endobiotica* (Gollerbach and Sedova 1974)—are good examples of this type of association. In these symbioses, despite their more profound relationship, some features of a physiological and morphological unity are missing: the integrity of a symbiotic complex and the coordination of the constituent components are not manifest.

An association in which one of the members is transformed in such a way as to become a link in the metabolic process of the other partner is a significantly more advanced symbiosis in terms of the extent of the relationship between the symbionts. An example of such a symbiosis exists between the mastigotes of the order Hypermastigida [Protoctista: Phylum Zoomastigina, Class Hypermastigida] and the termites, whose intestines these protozoa [protists] inhabit. Because they possess the enzymes necessary for breaking down the cellulose that termites eat, these mastigotes perform the most important function of the organism with which they coexist. The symbiotic relationship has truly become the organic link of parts of a single system. It is important to note that such a relationship led to the formation of a special mechanism that supports the existence and continuity of the entire system. At the base of this mechanism lies a particular form of instinctive behavior of termites—the instinct to lick (Dogel' 1962). The interrelations between ciliates from the family Ophryoscolecidae and the ruminants, whose intestines these symbionts inhabit, may serve as another example of the establishment of a physiological relationship (Polyansky and Strelkov 1935). [Editors' note: *Infusoria* is an obsolete name for ciliates and other protists, so when ciliates are referred to, we use the modern term, even though Famintsyn uses infusoria throughout.]

In certain cases, one of the organisms becomes a structural part of the organization of its partner (for example, the cytoplasmic inclusions of protists, Coelenterata, Turbellaria and other zoochlorellae, zooxanthellae, kappa-particles [in *Paramecium*], and so forth).

Finally, the ultimate stage of the synthesis of organisms includes associations that are bound so completely that it is possible to relate them to the various stages of the formation of a new complex organism, the life of which is regulated by a single physiological mechanism and facilitated by newly arising organs (for example, lichens).

The revelation of a series of successive stages of transition from loose associations, the components of which are bound by community relations, to increasingly profound and closely connected associations and, later, to a union of biological systems into an organism with physiological bonds between the symbionts is the basis for proposing evolution by means of symbiosis (symbiogenesis). Such a process is possible because each stage of the unification of heterogeneous organisms is a form of adaptation to the environment and is biologically beneficial in the struggle for existence. The driving force of such a process must be acknowledged to be group selection, which leads to the increase of the degree of integration of the symbiotic system.

The comparative analysis of changes in the organization of a series of associations within the limits of a definite group provides a factual basis for acknowledging the reality of the process of symbiogenesis. At present, the data necessary to construct a complete series of transitional forms has not yet been obtained. However, the situation has changed substantially during the half century since V. N. Lyubimenko wrote:

> On the basis of what is currently known in science about the symbiosis of plant organisms, we must admit the quite profound significance of this factor in the origin of complexes in which the independence of those organisms entering into the composition of a biological unit is little by little limited and even lost. Nonetheless, it must be recognized that, at present, we do not know such forms of symbiosis that could represent to us the gradual transition from unstable mutualism [the community form of the interaction of organisms—L. Kh.] to the complete merging of the symbionts into a single biological unit, which approaches the modern understanding of what an organism is. (1916, 183)

At present, the best studied model of the process of symbiogenesis in evolution is within the limits of the group of gelatinous lichens (Elenkin 1912, 1922a, 1922b, 1922c; Gollerbach 1928, 1930; Danilov 1929; Gollerbach and Sedova 1974). Crustose lichens were first used in research on symbiogenesis by Elenkin. He showed that the evolutionary transition from relations between a fungus and an alga as components of a community to a qualitatively distinct status (the organism lichen) is distinguished by the constant growth and refinement of the bond between these forms. If "the symbiosis of the alga with a fungus had stopped at the first stages of the evolution of those interrelations between the components that are observed in countless cases of symbiosis in the organic world," then the formation of a lichen as a single whole in the biological and systematic sense would be impossible (Elenkin 1975, 38).

The group of gelatinous lichens is especially interesting in that it includes forms that have advanced to various stages of the relation between the partners. Comparative morphological studies were made of the transformation of the external appearance of the thallus and the character of the interrelations of the components within this group as successive stages of the integration of the cyanobacteria of the genus *Nostoc* with the hyphae of the fungus, which is extracellularly endobiotic in the sheaths of these algae. Very often the relationship between these organisms resembles an accidental contact following the introduction of the fungus into the sheath of the alga, and therefore it is difficult to establish the exact emergence of a true gelatinous lichen. In these cases, the interrelations of the organisms do not differ essentially from the facultative interactions of populations of fungi and algae as components of an ecosystem. It would be more correct to define such an interrelationship as a "potential" lichen. This idea was expressed even more clearly by Elenkin during his analysis of the evolution of heteromeric lichens. At the first stages of their evolution, the relation between the green algae and the filaments of a saprophytic fungus form just like those between the components in any edaphic [soil] or surface microbial community (for more detail, see: ibid., 32–34).

Among those forms of gelatinous lichens that gradually develop more complex relations between their partners, the most primitive type is the lichen *Leptogium issatschenkoi*, described by Elenkin (1922a) and studied by M. M. Gollerbach (1930). In this species the relationship between the components is so weak that it [the fungus] has essentially no influence on the formation of the morphological appearance of the symbiotic thallus. The thallus of this lichen retains the form of the flat colonies, characteristic of several modified forms of *Nostoc commune*, one of which takes part in the formation of this lichen. The fungus endobiotic in the *Nostoc* gelatinous sheath causes no change in the alga's structure. It is also remarkable that, like the free-living cyanobacteria, the cells of this *Nostoc* form chainlike trichomes. The soredia of this species of lichen are quite imperfect: They develop in the thick of the thallus by the isolation of the sheath around several trichomes of *Nostoc*, which, continuing to grow, form isolated colonies. The frequent dissociation and formation of this lichen also demonstrate its primitiveness.

A somewhat more developed (but still primitive) type is the underwater gelatinous lichen *Collema ramenskii* Elenk. (Gollerbach 1928). In this case, despite the fact that *Nostoc zetterstedtii* remains the predominant component of the partnership, the influence of the fungal component is strong. The strengthening of the bond between the members of this symbiosis appears in the external organization of the thallus of the lichen. Instead of taking on the form of a lobed sphere, which is characteristic of colonies of this alga, the thallus appears as a flat plate, which has been deeply cut into narrow, branching blades with edges that are turned up and down. Under this influence, the disorganization of the trichomes proceeds: *Nostoc* cells become isolated from one another and appear as individual spheres assembled in agglomerations especially typical along the periphery of the thallus.

The homeomeric lichen *Synalissa* (*Pyrenopsis*) *conferta* can serve as the next in our series of lichens with gradually increasing degrees of the integration of their components. This species has established a closer connection because of the introduction of the filaments of a mycobiont into the thick sheaths of the cyano-

Symbiogenesis and Evolutionary Theory

bacterium of the genus *Gloeocapsa* and the attachment of these filaments to the cyanobacterial wall protoplast. The fungus is an endosaprophyte.

Saccomorpha arenicola, somewhat more advanced in terms of thallus formation, is still distinguished by the primitive type of relation between its components. The black, scummy thallus of this lichen consists of the septate branching hyphae of the fungus and dense, dark brown, sclerotized bodies that surround the filaments of the cyanobacterium of the genus *Stigonema.* However, the interrelations of the components are still primitive, far from a condition of prolonged symbiotrophy. According to Elenkin (1912), this is a case of a sharply expressed fungal parasitism leading to the disorganization and death of the alga. To maintain its existence, this fungal lichen must continuously acquire new individuals of *Stigonema.*

A more complex relationship between symbionts has been observed in the lichen *Pseudoperitheca murmanica* (Elenkin 1922b). This lichen, characterized by a muted form of parasitism, forms specific "nourishment" centers within which the algae are consumed by the fungus. The disorganization and death of the cells of the alga do not take place outside of these nourishment centers. This symbiotic relationship is the manifestation of yet another step toward the establishment of a more perfect relationship between the components—that is, the establishment of prolonged symbiotrophy.

Thus, the comparative study of a series of symbioses among gelatinous lichens shows well that the origin of lichens is a complex evolutionary process, in which the strengthening and complexification of the relation between the constituent components play leading roles. This process is reflected in the structural transformations of the lichenous thallus and the changes in the way symbionts nourish themselves.

The results of a study of the transformations among some endocyanoses [heterotrophs bearing cyanelles, that is, endosymbiotic modified cyanobacteria] (*Geosiphon pyriforme, Glaucocystis nostochinearum, Cyanophora paradoxa*), although somewhat less convincing, should be added to the sum of data that attest to this process. These data are reviewed in detail in chapter 7.

In conclusion, in the process of evolution the extent of the relationship between organisms in a symbiosis can vary. In the evolution of a majority of ecosystems, these changes always take place at the community level. In some cases, however, the evolutionary transformations of the symbiotic relations proceed so far and the integration of the components becomes so profound that the community relations develop into physiological ones, and the components become parts (tissues, organs) of a new, complex individual. The evolutionary change of community relations, resulting in the formation of a new complex organism or "symbioorgan" from the components of a symbiosis, can be labeled symbiogenesis in the narrow sense of the word.

Stages of Development of the Symbiogenesis Concept

Investigation of the relevant material allows us to distinguish four stages in the development of the concept of symbiogenesis [table 1].

The first stage occurred from the end of the 1860s to the period from 1905 to 1907 when preparatory experimental and theoretical research was undertaken. This stage was characterized by many years of experimental research on symbiosis and by the evolutionary and theoretical works of Famintsyn, which were based on the experiments of these years, as well as by the works of K. S. Merezhkovsky, in which were accumulated the materials that would serve as a basis for formulating the hypothesis of symbiogenesis.

The second stage, from 1905 to the beginning of the 1920s, was characterized by the elaboration of factual and logical arguments supporting the hypothesis of symbiogenesis, the first complete elaboration of the hypothesis itself, and the first systematic criticism of it.

The most important feature of the first two stages in this development is the fact that symbiosis was viewed as an evolutionary mechanism, in addition to

TABLE 1. History of Symbiogenesis.

STAGES	DATES	MAJOR CONTRIBUTORS	COMMENTS
1	1860s–1907	Famintsyn Merezhkovsky	Isolated experimental and theoretical work
2	1905–early 1920s	Famintsyn Merezhkovsky	Formal statements and criticism of symbiogenesis theory
3	1920s–end 1930s	Kozo-Polyansky	Incorporation of concepts of natural selection into symbiogenesis theory
	1940s–1950s	None	—
4	1960s–present	Many	Biochemical molecular biological revival

Symbiogenesis and Evolutionary Theory

those mechanisms elucidated by Charles Darwin, facilitating a new approach to the explanation of the gradual complexification of organisms in the process of evolution.

The third stage (from the beginning of the 1920s to the end of the 1930s) was characterized by the placement of the principle of natural selection at the foundation of the study of evolution by symbiosis (the work of B. M. Kozo-Polyansky). In these years, the most important aspects of symbiogenesis were reviewed critically by representatives of many areas of botany, and as a result, the foundations laid for trends in future research.

The ensuing two decades, especially the 1950s, are notable as a distinct pause in the elaboration of this problem. The beginning of the fourth stage, the current one, dates from the 1960s when interest was revived and an effort made to reconsider the significance of symbiogenesis within the modern theory of evolution on the basis of new facts. This book reviews the concept of symbiogenesis through the first three stages of its development and characterizes its current status.

E

2

EARLY CONCEPTS OF THE IMPORTANCE OF SYMBIOSIS IN EVOLUTION: THE WORKS OF FAMINTSYN

The experiments and ideas of Andrey Sergeevich Famintsyn (1835–1918) regarding symbiosis would be unclear without a general discussion of his works and opinions. Therefore, I briefly consider the extent of his scientific interests and his approach to the evolution of life.

Famintsyn was particularly prolific in plant physiology (Borodin 1919; Elenkin 1921a; Senchenkova 1960, 1973, 1974; Strogonov 1974; Genkel' 1974; Manoylenko 1974). His works—*An Experiment of Chemical and Physiological Research on the Ripening of Grapes* (1861), *The Effects of Light on Algae and Similar Organisms* (1866), and *Metabolism and the Transformation of Energy in Plants* (1883)—all of which contained many new ideas, were valuable contributions to the literature. Famintsyn's scientific organizational activity also played an important role in the history of science. The first Russian laboratory for research in plant physiology was created under his leadership in the Academy of Sciences (now the K. A. Timiryazev Institute of Plant Physiology of the Academy of Sciences of the USSR). Famintsyn, an outstanding teacher, founded the Petersburg School of botanists and physiologists. His students included I. P. Borodin, O. V. Baranetsky, D. I. Ivanovsky, V. V. Polovtsev, [and] D. N. Helyubov, among others.

Famintsyn's evolutionary studies proceeded in several directions, determined by both his understanding of Darwin's evolutionary ideas and his own experimental work in plant physiology (see Manoylenko and Khakhina 1974).

Famintsyn formed his ideas about symbiogenesis in evolution at the beginning of this century, presenting them most completely in his work *On the Role of Symbiosis in the Evolution of Organisms* (Famintsyn 1907a). He asserted that the increasing complexity of the organization and functions of organisms during the process of evolution may occur not only through the differentiation of simpler, early forms, but also on the basis "of the symbiotic unification of independent organisms into a living unit of a higher order" (Famintsyn 1907b, 143). This symbiotic unification could help attain "a possible method of the synthesis of complex life forms from more simple ones" (Famintsyn 1907a, 11).

First, Famintsyn attempted to outline those stages of phylogenesis in which

17

Andrey Sergeevich Famintsyn (1835–1918)

symbiosis could play an important role in evolution. The idea regarding the participation of symbiosis in evolution could be fruitful in the study of the origin of life on Earth and its subsequent development (Famintsyn 1918, 282). He proposed that plant and animal cells be viewed as "symbiotic complexes." The cell that contains the most important organelles as a result of selection (what we would call today the eukaryotic cell) arose symbiotically. This exceptionally bold idea, which seemed for a long time to be mere fantasy, deserves consideration and serious discussion in the current scientific literature. Famintsyn further asserted that the origin of one of the major taxa of plants, lichens, attests especially vividly to the effects of symbiosis as a factor of evolution. It was precisely in the symbiotic nature of lichens that Famintsyn saw the strictest proof of the role of symbiosis in evolution. Lichens, Famintsyn wrote, are of particular interest in that they are "the first directly observed case of the origin of a more complex plant form through the unification and interaction of simpler forms" (1907a, 11): that is, they are "an incontestable case of the participation of symbiosis in the evolution of organisms (1918, 281).

In his last and still unpublished work, *On the Role of Symbiosis in the Evolution of Organisms: A Preliminary Report on the Results of my Recent Work* (dated 1918),* Famintsyn summed up many years of deliberation on the significance of his ideas for the subsequent development of biology. He wrote: "Modern biology is threatened in the immediate future with the substitution of its most important principles by completely different ones and by complete restructuring on an entirely new basis. . . . I will list several examples; (1) Many organisms, now considered to be simple life forms, . . . will turn out to be based on two or more simple entities which, upon separation, continue to live independent lives but which, upon a change in the conditions of the external environment, are able to reunite and live a common life; (2) . . . The plant cell will be divided into several independent organisms or, in other words, it will be explained that the plant cell (like the animal cell) is a symbiotic complex; (3) . . . The role of cytoplasm in the life of the cell is very modest and does not have any special significance" (ibid.).

In substantiating the evolutionary significance of symbiosis, Famintsyn attempted to determine the primary conditions necessary for the stable existence of organisms that originated symbiotically. One necessary condition is the mutual transfer of the symbionts, or some of their parts or rudiments, from one generation to another (Famintsyn 1907a, 10). The search for a specialized mechanism providing offspring the ability to transmit genes from their parental symbiotic organisms today remains the most important, and as yet unresolved, genetic aspect of the entire concept of symbiogenesis.

*Archives of the Academy of Sciences of the USSR, Fund 39 / 1, unit 39, pp. 2–3.

Early Concepts and Famintsyn

There are two points of view of the causes that led Famintsyn to his conclusion about symbiogenesis. Borodin suggested that the results of experimental work investigating the nature of lichens were the major cause of Famintsyn's ideas on the significance of symbiosis in evolution. Famintsyn's discovery of zoospores in lichens "gave rise for the first time to the cult of the idea of symbiosis to which he—with amazing persistence—almost completely devoted the last period of his life" (Borodin 1919, 137). Elenkin expressed a different point of view. He believed that Famintsyn conceived symbiogenesis long before his discovery and intentionally chose lichens from a number of other possible subjects to prove his idea (Elenkin 1921a). Despite their disagreement, both of these opinions seem just, as they have been presented to us, because the factual material that Famintsyn attained in his work investigating the nature of lichens (without regard to whether he attained it before or after the origin of his idea of the significance of symbiosis) is only one of the prerequisites necessary for the origin of his entire concept.

Famintsyn's attitude toward Darwin's studies forms no less important a precondition for the creation of the symbiosis concept—that is, his striving to rethink critically and supplement the explanation of the causes of evolution from simple to complex forms (Senchenkova 1960; Khakhina 1973b, 1973c).

Famintsyn began accumulating material and developing ideas about the role of symbiosis in evolution early in his career. "Beginning in 1868, I constantly strove to extract from green plants a simpler organism that corresponds to the gonidia of lichens, supplied by the chloroplast and capable of living and reproducing outside the plant," he wrote much later (Famintsyn 1907a, 4). It was noted long ago that the concept of symbiosis as a factor of evolution goes back to Famintsyn's early experimental research analyzing the nature of lichens and the phenomenon of symbiosis in algae and other protists (Borodin 1919; Elenkin 1921a; and others). However, currently there is no work that illuminates this side of Famintsyn's work in detail. He conducted experiments in several different areas, including research on the formation of zoospores in different species of lichens, the study of symbiosis in several species of invertebrates and algae, observations of the behavior of chloroplasts in the seeds and shoots of sunflowers, the extraction of chloroplasts from plant cells, and attempted their artificial cultivation. I now turn to an analysis of these works.

Experimental Research on Lichens: Famintsyn's Interpretation of Symbiosis

The success of the works that exposed the symbiotic nature of lichens and that proved to be such an important foundation for the concept of symbiogenesis was conditioned by Famintsyn's approach as a physiologist to the study of this group of plants. The investigation of the nature of chloroplasts, their origin, and their functional role in the lives of plants facilitated an understanding of photosynthesis. The study of the green cells became the major subject of Famintsyn's research: he considered green cells of diverse genera of lichens such as *Evernia, Parmelia,* and *Cladonia* to be the most suitable subjects for his purposes.[†]

Famintsyn and Baranetsky stumbled across a curious phenomenon: only some gonidia were attached to the filaments of the medulla of the thallus, the majority lay freely between the hyphae, and when a transverse section was made across the thallus, they fell out easily, resembling single-celled algae. Precisely this accidental discovery led to the start of special research on the culture of gonidia outside the lichen thallus.

In their work, *A History of the Development of Gonidia and the Formation of Zoospores in Lichens,* Famintsyn and Baranetsky achieved the cultivation of algae outside of the lichen thallus by several methods. The most convenient method proved to be to soak sections that had been placed on the bark of spruce and lindens under running water. Under these conditions, the mycelial tissue of the lichen dissolved, and the gonidia separated easily, remaining "absolutely fresh and healthy." The gonidia reproduced, forming zoospores that are typical of many algae (Famintsyn and Baranetsky 1867).[§] From these experiments, the authors concluded that gonidia in this state "completely resemble the form *Cystococcus* described by Nageli." Studies of the development of *Cystococcus* have revealed how similar they are to the free gonidia of lichens.

Gonidia outside of the thallus of lichens undergo a change that results in the fact that most of them "begin internally to form zoospores" (ibid., 3–4). Famintsyn and Baranetsky drew an incorrect conclusion in addition to their correct ones. They thought that algae from the genus *Cystococcus* should be regarded as a stage in the development of the lichen (Oksner 1974, 37).

While demonstrating the possible development of gonidial zoospores in lichens, Famintsyn and Baranetsky strove to refute objection to the trustworthiness of their data. The purity of the gonidia, the high rate of zoospore formation in cultures of gonidia, the fact that, in all the experiments, the formation of zoospores occurred (on average) one month after germination—all of these were convincing arguments that eliminated all doubt that the formation of zoospores

[†]In the 1860s, the green cells of lichens [phycobionts, either chlorophytes or cyanobacteria] were considered to be organs of reproduction and were therefore called *gonidia.*

[§]Famintsyn later made an interesting remark regarding the degree to which each of the authors took part in this work: "In view of the fact that I offered Baranetsky the chance to contribute to this work, I considered it my duty to publish the results of our work in both our names despite the fact that the research methodology, the results, and the text belong exclusively to me" (Famintsyn 1914, 432).

did not come from the gonidia in these experiments. (The experiments were conducted with *Cladonia* sp. and *Evernia furfuracea*.)

The general conclusions resulting from this research were: green cells of lichens can exist for a prolonged period in culture (outside the body of the lichen); like algae and fungi, green cells can develop zoospores; gonidia are similar to single-celled algae (particularly to representatives of the genus *Cystococcus*) in structure and reproduction; and analogous results can be obtained from other species of lichens. At this point, lichens were still not classified as complex organisms, and no suggestion had been made as to how to form this taxon. The facts, however, were so thorough that the discovery of the true nature of lichens was only a step away. The most important factual basis for symbiogenesis was essentially created by this work. This is how Famintsyn himself evaluated the significance of this research.

Later, while studying the process of the formation of *Nostoc* on sections of the homomeric lichens *Collema pulposum* (*Collema tenax*) and *Peltigera canina* and comparing the free gonidia of these lichens with the algae *Nostoc* and *Polycoccus*, Baranetsky established that the gonidia of these lichens are "physiologically independent organisms"; capable of living "independently of the colorless tissue of the thallus and outside it as independent organisms," the free-living "gonidia of the lichens *Physica, Evernia* and *Cladonia* form zoospores, but the gonidia of *Peltigera* become dormant" (Baranetsky 1868, 56–57).

Many years later, near the end of his life, Famintsyn turned once again to the question of the formation of zoospores in lichens. In a special article, "On the Problem of Zoospores in Lichens," he attempted to refute the objections to his 1867 work made by M. Beyerinck (1890) and R. Shoda (1913). These authors expressed doubts about the correctness of the conclusions of Famintsyn and Baranetsky on the ground that there was no proof in their work that algae that had been observed producing zoospores were really the gonidia of lichens. Famintsyn wrote that "an important fact has escaped the attention of the critics. The formation and emission of zoospores were observed in green cells that were inoculated with pieces of the hyphae of lichens—that is, undoubtedly, in the gonidia" (1914, 431).

Together with his student V. Serk, Famintsyn shortly thereafter completed an experimental work that was supposed to dispel definitively the doubts concerning the authenticity of the results achieved in 1867. As in the previous experiments, but with different species of lichens (*Cladonia alpestris* and *Usnea florida*), zoospores were formed in the gonidia. They [Famintsyn and Serk] successfully extracted zoospores from gonidia that were intertwined with the remnants of the mycelial tissue of lichens and "fused with hyphae, which remained intact" (Famintsyn and Serk 1915, 1208). I now turn to the fundamental consequences of the discovery that Famintsyn and Baranetsky made in 1867.

Early Concepts and Famintsyn

S. Schwendener, guided by this discovery as well as by his own observations, proposed that gonidia were colonies of once free-living single-celled algae which gradually became entwined by the hyphae of a parasitic fungus (Schwendener 1869) and reached his own conclusion about the complex character of lichens. Famintsyn greeted this point of view with some caution. Thus, in 1870 he showed in his "Report on Schwendener's Work on Lichens" that the data collected by Schwendener could not resolve the problem of the nature of lichens. Only after some thirty-seven years had passed, while substantiating his position on the role of the synthesis of forms in the evolution of organisms, did Famintsyn write: "Schwendener's discovery that lichens appear to be constructions of fungi and algae constitutes his merit" (1907a, 3). K. A. Timiryazev especially valued the discovery of the complex nature of lichens (1888). In a lecture at the Polytechnic Museum, he stated: "Some botanists still cannot wake from the impression evoked by this startling discovery and prefer to close their eyes to the obvious. . . . If I am not mistaken, this curious subject has hardly been mentioned in our popular literature; nevertheless, it must be considered one of the most striking and unexpected discoveries of biological science of the last quarter century (Timiryazev 1937, 295).

As is well known, the phenomenon of the association of two heterogeneous organisms was reflected in the concept of symbiosis proposed by A. de Bary. After de Bary, Famintsyn also accepted the term *symbiosis*. However, in contrast to de Bary, Famintsyn understood symbiosis only as the mutually beneficial coexistence of heterogeneous organisms. Thus, for example, he wrote that one could distinguish cases of parasitism and symbiosis by the form of the interrelations of the organisms in an association (Famintsyn 1912b). In his last work, *What Are Lichens?* Famintsyn argued in favor of a mutualistic theory that excludes all other types of association from the definition of symbiosis. Famintsyn wrote that Schwendener "came to the conclusion that all associations are parasitic. Botanists throughout the world accept this theory, except for three individuals: Professor Elfving, Möller, and I."[‖] The expression "symbiotic association," which Famintsyn used, may serve as confirmation that he equated "symbiosis in general" with "mutualistic symbiosis." In this way, all of his statements attest to the fact that he seems to have narrowed the extent of the meaning of symbiosis in comparison with de Bary's interpretation of it. However, an analysis of Famintsyn's work clearly shows that he inserted an entirely new meaning into the concept of symbiosis. Neither de Bary, nor any biologist before Famintsyn, when defining the concept of symbiosis, proceeded from the possibility that the phenomenon of association is evolutionarily significant or even has adaptive value.

Famintsyn's position was distinguished by its clearly expressed evolutionary

‖Archive of the Academy of Sciences of the USSR, Leningrad branch, unit 39, p. 1.

approach. He viewed symbiosis as a special means for the evolutionary development of organisms, "as a means for the building of organisms with a more complex organization from several simple organisms" (Famintsyn 1916, 4). Symbiosis is a synthesis of organic forms in which association allows the creation of a more complex living entity (Famintsyn 1907a, 11). Famintsyn even introduced the term *formative symbiosis,* which in his opinion most fully and exactly reflected the meaning he intended for the term symbiosis. He explained: "In my research, I have in mind only formative symbiosis, in which the association of two or more symbionts forms a more complex organism" (1912b, 708). Lichens represent the clearest example of formative symbiosis.

Famintsyn was thus only interested in those cases of organisms that had been long and completely joined in association, that is, organisms of distinctive origin but profoundly integrated.

Experimental Research on Animal Symbioses

Famintsyn's research on the nature of the yellow and green bodies in several protists and invertebrates is related to those experimental works that, in addition to those surveyed above, served as an important basis for his views on the role of symbiosis in evolution (Famintsyn 1889a, 1891).

L. S. Tsenkovsky first showed that the yellow cells of radiolaria remain alive for several weeks after the radiolaria die, carrying out amoeboid movements, growing, and reproducing by cell division. The question of the structure of the green and yellow cells of animals, their function, and their ability to attain an independent existence—all of which penetrate to the essence of their nature—then continued to interest many scientists. The data that had been accumulated at that time were extremely contradictory. Thus, some scientists (Haeckel, P. Geddes, Sallit, Lancaster) considered these cells to be formed endogenously, originating by differentiation of the cytoplasm of animal cells. Others (Geza, G. Ents, K. Brandt, and later Danzhar and Beyerinck) saw in them independent forms of an algal nature, which had penetrated the cells from without and settled there symbiotically. Famintsyn (1889a), concentrating on the structure of the yellow cells (zooxanthella) of radiolaria and their role in the nourishment of their protist hosts, thought he could decisively answer the question of their algal nature. Although their structure was described in precise detail by K. Brandt in 1883, his opinions about their role in the nourishment of radiolaria, published in a monograph in 1885, seemed unjustified to Famintsyn. The results of Famintsyn's own experiments enabled him to resolve the question as follows: "The major role of the yellow cells consists in the fact that, like other algae, they are able to grow, to multiply, and to build their bodies from inorganic matter. During

ЗАПИСКИ ИМПЕРАТОРСКОЙ АКАДЕМІИ НАУКЪ.

MEMOIRES

DE L'ACADÉMIE IMPÉRIALE DES SCIENCES DE ST.-PETERSBOURG.

VIII° SERIE.

ПО ФИЗИКО-МАТЕМАТИЧЕСКОМУ ОТДѢЛЕНІЮ.   CLASSE PHYSICO-MATHÉMATIQUE.

Томъ XX. № 3.   Volume XX. № 3.

ТРУДЫ БОТАНИЧЕСКОЙ ЛАБОРАТОРІИ ИМПЕРАТОРСКОЙ АКАДЕМІИ НАУКЪ.

№ 9.

О РОЛИ СИМБІОЗА

ВЪ ЭВОЛЮЦІИ ОРГАНИЗМОВЪ.

— ·—

А. С. Фаминцынъ.

—

(Доложено въ засѣданіи Физико-Математическаго Отдѣленія 25 октября (8 ноября) 1906 г.).

С.-ПЕТЕРБУРГЪ.   1907.   ST.-PÉTERSBOURG.

The role of symbiosis in the evolution of organisms. *Transactions of the Academy of Sciences, series 8, Physical-Mathematical Division*, vol. 20, no. 3. St. Petersburg, 1907. (This illustration has been added to the English edition at the request of the author.)

25

this, they can serve as food for the radiolaria and support their lives during times of food shortage" (ibid., 32). Afterward, he researched the morphological traits, physiological features, and the ability of the green cells to maintain an independent existence in the ciliates *Stentor, Paramecium, Vorticella,* and *Stylonychia,* and in the sponge *Spongia fluviatilis* (Famintsyn 1891). His results established that the zoochlorellae of ciliates possess a cell nucleus, membrane, and a green chromatophore [chloroplast] that are clearly displayed during cell division, and that they [the zoochlorellae] are able to divide inside the ciliates. All the data on the structure of the green cells of the ciliates indicated that, in terms of their morphological features, they are typical single-celled algae (ibid., 6). The final judgment about the independence of these cells, however, can be obtained only by separating them from the ciliates and cultivating them artificially. This goal was achieved by using a pure culture of individuals isolated under a microscope: the cells of the zoochlorellae were not only separated from the ciliates but, over the course of several weeks, survived in culture, showing "the energetic growth and division" of the individuals, just as in the body of the ciliates (Famintsyn 1891, 12). The zoochlorellae were thus finally established to be symbiotic, single-celled algae that function as oxygen suppliers and food for the host cells. The symbionts were taxonomically identical to the free-living alga *Chlorella vulgaris.*

Famintsyn's works served as the transitional step to the study of the green inclusions of plant cells, chloroplasts. These works showed clearly that the extent of the evolutionary advancement of the relationship between symbiotic organisms could vary greatly.

Studies on Chloroplasts

After collecting experimental proof of the symbiotic nature of the green and yellow cells of ciliates and animals, Famintsyn began an investigation of the structural and physiological properties of chloroplasts, as well as a search for methods of isolating them from the cell and conditions for culturing them. "It seemed desirable to me to continue the work I had begun and to try to determine experimentally," Famintsyn asked whether "the chlorophyllous bodies of the simpler organisms containing them display a still greater independence than that which has been recognized in them and could they be capable of continuing to live and to reproduce outside the cell" (1907a, 4).

Famintsyn's article "On the Fate of Bodies of Chlorophyll in Seeds and Seedlings" (1893) was the beginning of this focus of his work. He considered the following question: Do the chloroplasts of seedlings form from the colorless chloroplasts found within the mature seed, or do they arise anew from the cytoplasm of the cells in the seed? The first point of view was supported by A. F.

W. Schimper and A. Meyer; the second was expressed by J. Sack, G. Haberlandt, K. Mikosh, and E. Beltsung. Additional data were necessary. These data were all the more important because they could confirm the hypothesis of plastid continuity in plant ontogenesis and thereby solve a more general problem—the problem of their phylogenetic origin. The seeds of the sunflower were chosen as the object of this investigation. The data obtained confirmed that "first, the chromatophores are found in the form of colorless leucoplasts in the mature seed, and that, second, the chromatophores of the seedlings form from them exclusively" (Famintsyn 1893, 13). In a more general form, this conclusion stated that "already prepared chromatophores in the form of colorless leucoplasts are found in the lobes of a mature, completely colorless seed of a sunflower and that, consequently, in the sunflower the new formation of chlorophyllous bodies directly from cytoplasm does not occur" (Famintsyn 1907a, 5).

Thus, although the concept of the plant cell as a symbiotic complex was formulated by Famintsyn only much later (when he wrote "is it not, though among the so-called protozoans, a symbiotic complex of two or more simple organisms?") (1912a, 1912b), the idea of the origin of the cell by symbiosis had arisen already in 1868. Therefore, it was natural that he would want to confirm this phenomenon with experimental proof. He wrote: "The next problem to be addressed is to find methods for culturing outside the cell those of its constituent parts that appear to be the centers of its life activity" (Famintsyn 1912b, 713). Most important was separating the chloroplasts from the cell and cultivating them artificially. There were two ways to solve this problem: by paralyzing the activity of one of the parts of the cell (for example, the part that contains the nucleus and the cytoplasm) and then supporting the activity of the other, or by cutting the cell into parts and searching for appropriate conditions for the growth of each part. The latter method was the most promising (Famintsyn 1912a, 57).

Famintsyn understood the entire methodological and, quite possibly, theoretical difficulty of solving this problem, the peculiarity of which lay in the fact that close and continuing symbiosis must have led to the specialization of the components within the cell, depriving them of the ability to exist independently. "The problem, both in terms of its novelty and the complete absence of other attempts to solve it, is so difficult that it is impossible to expect its conclusive resolution in the near future; it has been necessary, in taking up this work, to reconcile myself with this thought: to be satisfied with the elaboration of problems that would only outline the future path for solving this problem" (ibid., 65–66). For this reason he tried to obtain indirect proof, which would allow him to break down the plant cell itself and to isolate its plastids. The most important observations in this regard were obtained in experiments with *Vaucheria* sp. and the sea alga *Bryopsis muscosa* (Famintsyn 1912a). The presence within the cells of *Vaucheria* of two independent protoplasts—one with the chlorophyllous bodies and one that was colorless with bodies, microsomes, and vacuoles—was well substantiated, as Famintsyn thought, by the temporary changes of the distribution of the contents at the growth tip ("by the growth of vacuolar plasma with bodies into a green mass of the plasma with chlorophyllous bodies"); by the "migration" of the body into the exterior layer of peripheral plasma with the zoospore, which is still included in the zoosporangium; as well as by the "regrouping" of the chlorophyllous body with the cytoplasm surrounding them and that of the interior layer of plasma with the bodies, vacuoles, and other inclusions during the development of the sexual organs of *Vaucheria*.

Famintsyn not only collected the data available at that time that might serve as indirect arguments for the independence of chloroplasts, but (over the course of more than forty years) he experimented with the separation of chloroplasts, hoping to obtain direct proof of the symbiotic origin of the plant cell. The dissection of a cell of a filament of *Vaucheria* led only to its breaking apart into its constituent elements, which took on a spherical form. Observations of the constituents of *Bryopsis* noted "the structure and movement of the bodies of chlorophyll" and traced their "transformation" into spherical bodies similar to zoochlorellae (ibid., 63–64). This work did not yield the expected results. "I have not yet managed to achieve this result, and I am continuing to experiment in this area," Famintsyn wrote (1907a, 4). It was precisely the separation of the chloroplasts and their artificial cultivation that Famintsyn considered the most important proof of their symbiotic origin. At that time, however, no one could culture isolated chloroplasts. The difficulty was in finding a suitable artificial culture medium. Only in recent years have results been obtained (for more detail, see: Butenko 1971). The data of experiments in which isolated chloroplasts were injected into fibroblast cells of mice, conducted by the [German] American researcher Margit Nass (1969), are also of distinct interest. The introduction of the chloroplasts of the leaves of fresh spinach (*Spinacia oleracea*) and of the African violet (*Saintpaulias*) into mammalian cells was successful because of the phagocytic characteristics of the mice fibroblasts (L-cells). Chloroplasts were seen in L-cells over a period of five days: they divided, preserving to a significant extent their structural integrity, and they manifested an excited reaction to light and an ability to assimilate CO_2. Using an electron microscope and photochemical analysis, Nass showed that the "absorbed" chloroplasts preserve their structural and functional integrity over the span of several generations.

Famintsyn's experimental works were directed toward a single goal: to explain and model the process of the formation of the complex organism from simpler forms of various origins. In this regard, his most important work was his research on lichens, which, in the author's own opinion, provided "the first intelligible, strict verification of the construction of a complex organism from simpler ones" (Famintsyn 1907a, 3). Famintsyn was the first who demonstrated experimentally the significance of symbiosis as one of the factors in the evolutionary process.

All of Famintsyn's experimental work, which began in the mid-1860s and spanned more than half a century, was dedicated to proving the possibility of the evolutionary synthesis of complex organisms from simpler ones and must be considered among the first serious scientific foundations of symbiogenesis.

Famintsyn's Symbiogenesis

Both experimental and other sources of information formed the basis of Famintsyn's theory regarding the role of symbiosis in evolution, related to his understanding of the teaching of Charles Darwin. Famintsyn was one of the first Russian biologists who accepted Darwin's teachings and regarded them highly. As early as 1874, Famintsyn gave a speech at St. Petersburg University called "Darwin and His Significance for Biology," in which he described [Darwin's] teachings and emphasized their exceptional value for biology. The extraordinary significance of Darwin's findings was that he was the first one to demonstrate the mutability of organisms, putting an end to the notion of the constancy and immutability of species, which "sometimes exceeds the limits that have been accepted for clearly established species," and he also demonstrated how new species originate (Famintsyn 1874, 19).

Darwin's most important contribution, Famintsyn later asserted, was in being the first to explain strictly scientifically organic expedience, the development of that "marvelous harmony" that has been established between living entities and the nature surrounding them (1894, 135). If previously such problems as, for example, the correlation of an animal's color and its environment had been "beyond strictly scientific investigation," then now they came to be among "the category of problems amenable to a scientific approach" (Famintsyn 1889b, 633).

Famintsyn insisted that Darwinism had become the "point of reference for the studies of the materialists while simultaneously denying the scientific basis of the studies of the idealists" (ibid., 624) and had provided a method that made working out morphological problems possible, not from the position of the doctrine of purposefulness, but "from the perspective of the causality of phenomena" (ibid., 633). "The determining question in his [Darwin's—L. Kh.] research is the question 'from what?' not 'for what?' as it had been for de Bary and his followers. Darwin will always have indisputable merit in that he not only made proposals but was able to win broad support for similar research" (ibid.).

Famintsyn did his part to defend Darwinism. Although his work along these lines was not as clear and significant as, for example, the activity of Timiryazev, his defense of Darwin's theory during the 1880s, when attacks against it were fierce, had an enormous effect.

In 1889 Famintsyn took part in the discussion centered around N. Ya. Danilevsky's book *Darwinism*. This book, which attempted to refute Darwinism, along with the ideas of its defender N. N. Strakhov, met with a heated rebuff from Timiryazev. A polemic developed (see Raykov 1957). In the article "N. Ya. Danilevsky and Darwinism: Has Danilevsky Refuted Darwinism?" (1889), which appeared in the journal *The Herald of Europe* ("Vestnik Evropy"), Famintsyn expressed his attitude toward the discussion, toward the positions of Danilevsky and his opponents, and came to the defense of Darwin's doctrine. He declared that he had "always considered it my sacred obligation to defend the significance of Darwin and his teachings in science from nonsensical criticism, having accepted him as one of the greatest naturalists of our century" (Famintsyn 1889b, 627). In an attempt to clarify the polemic, Famintsyn set himself the task of providing a "thorough and impartial" analysis of Danilevsky's book. However, in reality, he avoided a critical analysis of Danilevsky's arguments, and Timiryazev, who himself had written a brilliant analysis of this book, justly reproached him for his uncritical assessment (1889). Famintsyn did not consider a detailed analysis to be expedient because Danilevsky's views were not original, and many of his arguments had already been "expressed in detail by his predecessors" (Famintsyn 1889b, 627), and those few that were the author's, in themselves, "had no decisive significance" (ibid., 639). Famintsyn's article, however, condemned the book *Darwinism* as a whole. He declared Danilevsky's opinion of the significance of Darwin's scientific work to be profoundly mistaken and bemoaned the appearance of a book that interpreted Darwin incorrectly as "an extremely regrettable, undesirable, and harmful phenomenon" (ibid., 643).

Although Timiryazev was interested in the extent to which the facts and arguments introduced by Danilevsky against Darwin were persuasive, Famintsyn's attention was attracted first of all by the motives that had impelled Danilevsky to come forward. In Famintsyn's opinion, the fundamental reason for this attack lay in the fact that Darwinism had become the "point of reference" or "the scientific basis" of the materialistic study of nature (ibid., 624). This critique of Danilevsky's anti-Darwinism was extremely convincing and precise.

Although recognizing the enormous significance of Darwin's works, however, Famintsyn found distinct shortcomings in Darwin's concept of evolution. The interpretation of these shortcomings and his efforts to correct them lay at the basis of the theoretical generalizations that formed Famintsyn's unique attitude toward Darwin's doctrine. Proceeding from them, he arrived at the idea of symbiosis as a mechanism of evolution.

Early Concepts and Famintsyn

Famintsyn was apparently one of the first Russian biologists to pose clearly the problem of the inequality of changes—that is, their inequality in terms of evolutionary results. He distinguished two types of evolutionary change: changes that secure the adaptation of organisms to external conditions without manifesting a change in the complexity of their structure, and changes that lead to an increased or decreased complexity of organization and that allow the organism "to move beyond its normal limits" (Famintsyn 1894, 136). He called the first type of change "changes of the plasticity of organisms" or Darwinian changes, and he called the second evolutionary changes. Proceeding from these concepts, Famintsyn arrived at the theory that two independent processes exist: one that proceeds horizontally and is associated with the best adaptation of the organism to the environment, and the other, which is evolution—that is, a process that leads to the development of complex organisms from simple forms (ibid., 136). Darwin's doctrine of natural selection provided a singularly plausible and completely satisfactory explanation of the causes of the formation of adaptations. As far as the causes of the evolutionary complication of forms is concerned, however, they "are not in the least elucidated by Darwin and remain, as before, an unresolved question" (ibid., 137). These opinions, at first glance, seem remarkably similar to the teleological opinions of K. Nägeli (1817–1891) regarding the existence of two unrelated alterations of traits (*organizational* and *adaptive*) and the causes behind them. In reality, this is not the case. By separating adaptive traits from organizational ones, Nägeli (1884) denied that natural selection plays a leading role in the evolution of adaptive traits and believed that they evolve as a result of the innate capacity of organisms to react expediently to external influences. He also explained the evolution of organizational traits by the action of the teleological principle—by the action of the law of "*striving for perfection.*" In contrast to Nageli, Famintsyn was convinced that natural selection of the individual changes of organisms was the driving force of the evolutionary process of adaptation. Thus, in terms of the basic question of evolutionary theory (the question of the causes of the formation of adaptations), Famintsyn's position was Darwinian and was directly contradictory to Nägeli's teleological convictions. As far as the causes that advance evolution are concerned, this question, in Famintsyn's opinion, demanded further clarification and research. His idea that Darwin's teaching did not provide an answer to the question of the methods of the development of complex forms from simple ones cannot be recognized as true. It is now known that the driving force of evolution, which is realized in all of its tendencies, including the tendency toward progress, is natural selection.[1] Famintsyn's

---

[1]A profound and thorough investigation of the question at hand is given by K. M. Zavadsky in his article "Research on the Driving Forces of Arogenesis" (1971).

judgment, however, was correct in one regard: he showed that Darwin himself had not managed to elaborate a theory of progress and that his understanding of this question was not sufficiently clear or consistent (Zavadsky 1958, 1967).

Darwin was vague in his application of the concept of "progress," which he interpreted inconsistently. On the one hand, he asserted that the action of natural selection always and inevitably leads to the perfection of organization since, because of natural selection, "each creature manifests a striving to become increasingly perfect with regard to the conditions which surround it" (Darwin 1939, 359). Thus, general progress is identified with specialization. On the other hand, Darwin expressed the opinion that natural selection does not presuppose mandatory progressive development (ibid., 361).

To fill this gap in Darwin's thought, Famintsyn tried to find an evolutionary factor that would satisfactorily explain the causes of development from the simple to the complex within the tendency of the perfection of organization.

Originally, in the works of the 1890s, Famintsyn (1890, 1894, 1898, 1899) identified this factor in the phenomena of "intelligibility," "the rationality of behavior," in "mental processes" that are supposedly present in all living creatures. He explained the similarity of his theories with Lamarck's views on the influence of the mind on phylogenetic development in the same way. "The similarity of my opinions with Lamarck's view of the influence of necessities and habit, of character, inclination, activity, and even of thoughts—that is, the mind—on organization and, consequently, on phylogenetic development as well," Famintsyn wrote, "has become very clear, particularly with regard to the following point: in the recognition of the active participation of the mind both in phylogenesis and in the formation of a final organization of animals and humans" (Famintsyn 1898, 172). These mistaken psycho-Lamarckian ideas were thoroughly and deservedly criticized by Timiryazev. At this point, in something of an aside, I return to a characterization of Famintsyn's general views on evolution.

That Famintsyn's views supported some Lamarckian ideas served as an approach to evaluating the entire system of his evolutionary views as anti-Darwinian. A close familiarity with his works, however, shows clearly that there are no grounds for attributing an anti-Darwinian tendency to his evolutionary views or for numbering this scientist among Russian anti-Darwinists. A detailed evaluation of Famintsyn's views does not contradict—but rather coincides with—the many statements and remarks made by Timiryazev regarding Famintsyn's activity. It should be emphasized that the majority of these remarks were sharply critical and were, in essence, directed against Famintsyn's phytopsychological ideas and his unfulfilled attempts to write a "thorough" and "impartial" review of Danilevsky's book *Darwinism*. This was a just and strictly scientific criticism of Famintsyn's mistaken theories; nonetheless, it did not contain even a hint of a characterization of Famintsyn as a representative of Russian anti-Darwinism.

However, even in his works from the 1890s Famintsyn allowed a totally different possibility for explaining the causes of the increasing complication of organization. Complex organisms might arise "through the unification of elementary organisms into colonies, and the transformation of the aggregate of them into an entity of a higher order . . . living, so to say, a sum of the lives of the many thousands of elementary organisms that constitute it" (Famintsyn 1890, 38). This thesis, expressed initially as if in passing, was later developed—clearly influenced by the results of his experimental works on symbiosis—and became his basic concept of symbiogenesis (Famintsyn 1907a, 1907b, 1912a, 1912b, 1916, 1918).

Famintsyn also connected his consideration and general evaluation of Darwin's doctrine with the extent to which his basic theories were proved by data. Since Darwin's doctrine at that moment was based only on indirect proofs, he asserted, it could be labeled only "a working hypothesis" or "a highly probable theory" (Famintsyn 1907a, 1). These statements, which now seem to be of only historical interest, led Famintsyn to the conviction that direct and proximal proof of evolution might be obtained on the basis of symbiosis. Thus, for example, a lichenous organism is better characterized as a living entity by its complexity of organization and by its physiological functions than by the components (fungus and alga) that constitute it. This was a direct proof of the evolutionary origin of complex organisms from simple ones in Famintsyn's opinion. "The synthesis of lichens," he asserted, "is the first directly observed fact of the origin of more complex plant forms through the unification and interaction of simpler ones, . . . the first irrefutable factual proof of the theory of the evolution of organisms" (ibid., 11).

Thus, it was precisely his unique interpretation of Darwin's concept, his desire to clarify it and to supplement it so that it could serve as an explanation of the progressive development of organisms, as well as his striving to produce empirical proof of evolution and thereby to put Darwin's theory on a firm foundation that caused the development of the symbiogenesis concept. Eventually, this concept came to be considered by Famintsyn as "a new scientific theory," capable of explaining "the phenomena of life on Earth and the basic laws of its further development" (Famintsyn 1918, 281).

In our opinion, Famintsyn must be acknowledged as the founder of the strict scientific approach to the verification of the symbiogenesis hypothesis and the author of a productive supplement to the explanation of the means for increasing the complexity of organization that had been set forth by Darwin.

M EREZHKOVSKY'S HYPOTHESIS OF SYMBIOGENESIS

3

In the first years of the twentieth century, Konstantin Ser-geevich Merezhkovsky (1855–1921) began to develop ideas similar to those of Famintsyn about the evolutionary significance of symbiosis. Merezhkovsky, like Famintsyn, argued that evolutionary transformation can occur by the integration of symbionts, two or more simple organisms differing in phylogenetic classification. Nevertheless, their approaches to these ideas differed.

Merezhkovsky first presented his views on the role of symbiosis in evolution in 1905. His article "The Nature and Origins of Chromatophores in the Plant Kingdom," preceded Famintsyn's work (1907) on the same topic. The correspondence between Famintsyn and Merezhkovsky, however, now housed at the Leningrad branch of the Archive of the Academy of Sciences of the USSR, shows that Merezhkovsky was interested in Famintsyn's work and in the experimental results on the nature of chloroplasts.*

On 7 May 1903, Merezhkovsky wrote to Famintsyn, "I am extremely inter-ested in the question of the significance of the chromatophoric nuclei (whether they are organs or independent organisms) and intend to work in this area. Knowing that you also have worked on this question and even, it seems, at-tempted to cultivate them in a controlled environment, I most humbly request that you send me a reprint of your work on this question or indicate where in your work this question is discussed."

Analysis of the correspondence between Merezhkovsky and Famintsyn sug-gests that, despite knowledge of each other's research and exchange of opinions, these two investigators each formulated similar evolutionary concepts indepen-dently of one other.

Merezhkovsky entered the annals of biology as a scholar, one of the first to elaborate a concept of the cell's symbiogenetic nature. According to his bold hypothesis, the fundamental cellular organelles originated through intracellular symbiosis. He especially persisted toward proving the role of symbiosis in the evolution of the green plastid of plant cells. The name Merezhkovsky is the first to be mentioned when the question of the origin and evolution of the photo-

*Leningrad Branch of the Academy of Sciences Archive, 39, 2, 18, 5.

34

Konstantin Sergeevich Merezhkovsky (1855–1921)

synthetic apparatus of plants is discussed (Famintsyn 1907; Lyubimenko 1916; Kozo-Polyansky 1924; Breslavets 1959; Ris and Plaut 1962; Kirk 1970; Raven 1970; Taylor 1970; Belozersky, Antonov, and Mednikov 1972; Svetaylo 1973; Senchenkova 1973; Takhtadzhyan 1973, and so on). Merezhkovsky's views on the significance of symbiosis laid the framework for phylogenetic systematics. His ideas attracted the attention of researchers (Smirnov 1952; Kusakin and Starobogatov 1973; Takhtadzhyan 1973) and are valued as having "substantial significance for understanding the evolution of higher taxonomical units of the organic world and the construction of its system" (Takhtadzhyan 1973, 22). What is incomprehensible is that, in certain research, any reference to Merezhkovsky's work is absent. For example, in the work of L. Margulis (Sagan 1967; Margulis 1970), expressly devoted to the origin of the eukaryotic cell and the history of that question, neither Merezhkovsky nor Famintsyn is mentioned.[†]

Merezhkovsky formulated the symbiogenetic hypothesis during the Kazan period of his career: from 1902, when he was appointed curator of the zoology library at Kazan University, until 1914, when he was dismissed from the university. In these years he formulated the fundamental tenets of the concept, suggested the term *symbiogenesis,* and collected and summarized data (Merezhkovsky 1905a, 1906, 1909a, 1909b, 1910). Preceding the Kazan period was a major stage of research on morphology, physiological peculiarities, and the geographical distribution of a number of species of protists, sponges, hydroids, and coral.

Initially, Merezhkovsky was interested in comparative morphological and zoogeographical questions; subsequently he gave even more attention to the distribution of pigments in animals. He researched these questions on broad comparative material (Merezhkovsky 1883). He was concerned with animal pigment even after he began research on algae in 1894. Over the course of many years, Merezhkovsky researched the morphology of chromatophores of diatoms—their structural features—which are covered in the monograph "On the Morphology of Diatom Algae" (1903) and in the work "Laws of the Intracellular Pigment" (1906).

Analysis of data gathered from studying the diatom pigments, in Merezhkovsky's words, led to his conviction that the nature of chromatophores in plants is completely unique: "They are considered special independent bodies, which are located in the cytoplasm but are in no way similar to it" (Merezhkovsky 1903, 128–29). Studying the diatom chromatophores "opened a kind of special, origi-

[†This oversight was brought to my attention by A. Takhtadzhyan at the 1975 Botanical Congress in Leningrad and remedied in Margulis 1981, 1992.—L. M.]

nal, and in the highest degree, curious world of phenomena; inside the cell of the diatoms we are confronted with some kind of apparently independent organisms that live in the cell like guests, developing according to their own laws, dividing themselves, dependent on it only in so far as organisms are in general dependent on their surrounding environment" (Merezhkovsky 1906, 5).

Observations of diatom algae led Merezhkovsky to reexamine the existing hypothesis about the nature of chromatophores, requiring a new explanation of their origin. The first attempt to summarize these data resulted in six rules or laws of pigment (Merezhkovsky 1906). A theoretical interpretation of natural laws from the evolutionary point of view, Merezhkovsky believed, had to proceed from the assumption that chromatophores were originally independent, free-living organisms, which in remote geological epochs had established themselves in colorless cells, forming a close symbiotic relationship with them. Chromatophores persist from generation to generation aided by gametes and spores (ibid., 350, 353). He formulated this view of the nature and origin of chromatophores completely independently of and "long before creating the theory of symbiogenesis," primarily based on asynchronous division of the cell and its chromatophores in several types of diatoms (ibid., 353).

These and others of his statements allow us to conclude that Merezhkovsky's research on diatom chromatophores prompted his hypothesis on the symbiotic origin of pigment carriers in plant cells. His personal observations, especially on the pigment bodies in animals and plants, were the main source of his views on the role of symbiosis in the origin of fundamental cell organelles.

Merezhkovsky suggested two relatively independent hypotheses for the general concept of symbiogenesis. The first was the view that plant and animal cells are complex combinations and symbiotic associations of once free-living and primitive living beings. This suggestion in turn arose from his supposition about the symbiotic nature of chromatophores and the nucleus[5] of the plant cell. The second hypothesis is known as *the theory of two plasms*. Both expressed general conclusions and despite their relatively independent character were related: this understanding of the symbiotic nature of chromatophores became a component of the theory.

We will examine each of these hypotheses in detail.

---

[5]Merezhkovsky suggested that the question of the nature and origin of the nucleus be examined. He wrote: "The entire part of my study about the origin of organisms, which concerns the nucleus, is the subject of a special article, in which the facts—which serve as a basis of those statements presented in brief form here—will be discussed [the book *The Theory of Two Plasmas*," is referred to here—L. Kh.]" (Merezhkovsky 1909a, 92). The author's intention, however, remained unrealized.

Chromatophores

Merezhkovsky's idea of the symbiotic origin of chromatophores began with his belief that the conviction that chromatophores arose from cytoplasmic differentiation and gradually assumed the special function of photosynthesis, though widely held, was itself no more than an unsubstantiated hypothesis (Merezhkovsky 1905a). This conviction, Merezhkovsky emphasized, came from an analogy with observations of the colorless parts of plants, which turn green after exposure to light. Precisely this fact lay at the root of the conclusion that carriers of chlorophyll—chlorophyll-containing bodies—appeared de novo each time. That this was an unsound assertion became apparent after Schimper's research; Schimper showed that plastids are formed by division and are transferred from generation to generation not only by vegetative but also by sexual reproduction—that is, he proved the continuity of plastids from generation to generation.

The continuity of plastids through the series of cell generations showed that they are never formed de novo through cytoplasmic differentiation. They arise only through the division of similar forms preceding them (ibid., 594). Accepting the continuity of chromatophores as an established fact, Merezhkovsky asserted that the widely held notion of their phylogenetic origin must be reexamined.

If one accepts this, then the suggestion that chromatophores are not "organs of the cell and never were" is the only possible assumption. They must be viewed as bodies that derive from outside the cell but that began a joint existence with it (Merezhkovsky 1905a, 596; 1910, 66). "Chromatophores," he wrote, "should be viewed not as organs or organelles that gradually differentiated in the cell, but as ones that were not previously included. Rather, they were independent organisms that never established associations in colorless, animal cells (amoebae and flagellates), having entered into a close symbiotic relationship with other cells and now being continuously transferred from generation to generation with the help of gametes and spores" (Merezhkovsky 1906, 350). For Merezhkovsky, this conclusion merely verified an existing fact. At the same time, the truth of the assumption that chromatophores are symbionts was a reliable basis for the theory of the symbiotic origin of plant cells in general.

The concept of the continuity of plastids—an argument to which Merezhkovsky gave great significance, is not now a generally accepted point of view. If for plants such as mosses (Sapegin 1913) and single-celled algae (Green 1964) chloroplast division is certain, then among green plants the propensity of plastids to divide, and particularly their behavior in sexual reproduction, demands more research. Two opinions exist concerning the origins of chloroplasts in cells of the meristem of vegetative organs and their transfer from one cell generation to another. One is traditional: plastids result from the division of previously existing plastids (Kozo-Polyansky 1924; Kirk 1970; Robertis, Novinsky, and Saes 1973; Senchenkova 1973; Zhukova 1975). According to the second view, the plastids of land plants lost the capacity for division; ontogenetically they develop de novo from submicroscopic cytoplasmic particles (Alexandrov 1950; Breslavets 1959, 1963) or are reconstructed from bodies that emerge from the nucleus (Mühlethaler and Bell, 1962). Recently, an attempt was made to resolve differences on the question of the continuity of plastids, proceeding from the differences in morphogenesis in the plastids of plants and algae. The question of the continuity of plastids through the sexual reproduction of plants demands more precise work (for more detail, see: Zhukova 1975).

Merezhkovsky considered the great self-sufficiency and functional independence of chromatophores from the nucleus of the cell to argue for their symbiotic nature. Observations made by G. Klebs and J. Reinke on functioning, reproducing plastids in those parts of cells lacking nuclei were convincing to Merezhkovsky. His data on diatom chromatophores also supported this phenomenon. Merezhkovsky thought that the independence of diatom chromatophores was most clearly seen by the absence of strict synchronicity between cell division and chromatophore division—appendix 3 in "Laws of the Intracellular Pigment" (1906). In *Navicula radiosa* and *N. anglica*, for example, the cell divides earlier than the plastids. In *N. gracilis* and *N. rhynchocephala*, plastids divide prior to any signs of cell division. In the family Surirellaceae, division of the plastid always precedes cell division. Merezhkovsky saw signs of plastid independence in the lack of correlation between the plane of cell division (constant) and that of the dividing plastid (variable). Comparing the growth of the diatom cell with the growth of its plastids showed no direct correlation, which also supported the concept of plastid independence.

Summarizing these arguments, Merezhkovsky wrote: "Chlorophyll bodies grow, are nourished, reproduce, synthesize proteins and carbohydrates, hand down their characteristics—all independent of the nucleus. In a word, they behave like independent organisms and therefore should be examined as such. They are symbionts, not organs" (1910, 67).

Data supporting the independence or, as it is now called, the autonomy of plastids, have been reviewed (Kirk 1970; Svetaylo 1973; Senchenkova 1973; Pinevich et al. 1974, and many others). Plastids possess significant functional and genetic autonomy (see chap. 7).

Merezhkovsky also compared chromatophores to typical endosymbionts (zoochlorellae and zooxanthellae), a serious argument in favor of his concept. Their resemblance seemed so compelling as to suggest the existence of a "complete analogy" between them (Merezhkovsky 1905a, 598; 1910, 67; 1920, 52–56). The analogy was especially clear in light of the similarity of morphological features and methods of their transfer from generation to generation. The major difference between chromatophores and zoochlorellae was the capacity of the latter to live outside the host cell, while all attempts to grow plastids in artificial environments failed. In Merezhkovsky's opinion, however, this could not diminish the analogy because the symbiosis of chromatophores and the first bacterial-type cells had arisen much earlier than the symbiosis of green and yellow algae with ciliates, hydras, and so forth.

Of all the arguments raised by Merezhkovsky in favor of the symbiotic origin of chloroplasts, the argument about the similarity between symbiotic green and yellow single-celled algae and chromatophores met the most criticism. Even Famintsyn reproached Merezhkovsky for the arbitrary declaration of a full analogy between chromatophores and zoochlorellae. He underscored that the author "seemingly forgets" about the fundamental difference between them: the presence of nuclei and membranes in zoochlorellae and the their absence in chromatophores (Famintsyn 1907a, 6). Gollerbach and T. L. Sedova were critical of Merezhkovsky's argument in a statement made in an article about symbiosis in algae. He "completely identified the symbionts in the animal cells of algae (zoochlorellae and zooxanthellae) with chloroplasts. Despite this, Merezhkovsky did not consider important differences (already well known at that time) between them as in the presence in zoochlorellae and zooxanthellae of a morphologically distinguishable nucleus and cell wall and the absence of these structures in chloroplasts" (Gollerbach and Sedova 1974, 1370). Although he mentioned resemblances between zoochlorellae and zooxanthellae and chromatophores, Merezhkovsky did not identify them. Apparently, it is more accurate to assume that he consciously overemphasized those aspects of the phenomena that might help prove his hypothesis. Certainly such a method cannot be recognized as irreproachable but rather is completely allowable in the presence of a logical conceptual basis.

Merezhkovsky attributed the great similarity of chloroplasts and cyanobacteria to the significance of symbiosis in the origin of chromatophores. These similar traits were so numerous that it was impossible to view them as coincidence and left no doubt of a direct evolutionary relation between them.

The most probable precursors of chromatophores are the primitive representatives of a type of cyanobacteria. Among currently living things there exist "organisms, which can be viewed as free-living chromatophores—for example, the lower forms of Cyanophyceae (for example, *Aphanocapsa* or *Microcystis*)" (Merezhkovsky 1905a, 599). This conclusion was based on Merezhkovsky's comparison between the blue-green cells and those of green algal chromatophores. Comparison of the volume, size, inner structure, and method of reproduction showed their great resemblance. Merezhkovsky considered the most essential difference between them to be the absence of a cell wall in chromatophores and its presence in blue-greens. He explained this difference by the prolonged residence of blue-greens as symbionts in the cells of plants, which corresponds well with the absence of a wall in green algae symbiotic in turbellarian worms.

The parallel made by Merezhkovsky between Cyanophyceae and green plastids seemed to Famintsyn completely arbitrary (1907a, 6). Famintsyn approached the solution of this question with exceptional caution, and he never expressed his opinion about which of the free-living and chlorophyll-possessing organisms might be chloroplast precursors. Only in the most general sense did he write about early forms that could have given rise by symbiosis to the plastids of the plant cell; these were two of the most simple organisms: one of them is green with chloroplasts, the other is a colorless, amoebalike form, consisting of cytoplasm and nucleus (Famintsyn 1907a, 1912a, 1912b). Merezhkovsky's assumption about symbiotic blue-greens as the only possible ancestors of chloroplasts did not enjoy the support of Kozo-Polyansky either (1924), who emphasized that Merezhkovsky had mistakenly equated the pyrenoid in chloroplasts of algae with the central body of blue-greens. Indeed, in showing the similarity of primitive forms of blue-greens to chloroplasts, Merezhkovsky erroneously identified two inclusions of differing natures: the central body (nucleoid) of the blue-greens and the pyrenoid in the plastids of other algae. Moreover, he suggested that taking into account the "evolution of plastids" was necessary in order to compare the structures of blue-greens with the most primitive form of chromatophores, where the pyrenoid is clearly present (Merezhkovsky 1910, 67). These propositions were subjected to valid criticism that is impossible to ignore (Gollerbach and Sedova 1974). At the same time, it is difficult to agree with their opinion that Merezhkovsky failed to establish a proper basis for the similarity between cyanelles and chloroplasts. In contemporary research, Merezhkovsky's idea that blue-greens are the possible ancestors of chloroplasts is receiving well-known corroboration. This information is included in chapter 7.

Thus, as early as 1905 Merezhkovsky had clearly and rigorously formulated a hypothesis on the symbiotic nature of the photosynthetic apparatus of the plant cell. It was based on an acknowledgement of the direct phylogenetic tie between blue-greens and chloroplasts and an assumption of hereditarily consolidated, intracellular symbiosis as a mechanism of their origin. Merezhkovsky's ideas had principal significance for the development of theoretical bases of evolutionary study as well: symbiosis was recognized as a special and essential factor in evolution.

The Theory of Two Plasms

We now examine the other basis of the concept of symbiogenesis. Merezhkovsky assumed a dual nature of the organic world: all life consists of two essentially different plasms. This idea was developed in a small book, "The Theory of Two Plasms as the Basis of Symbiogenesis: A New Study on the Origins of Organisms" (1909). Radically dissimilar in their physical-chemical and functional peculiarities—two plasms make up all living organisms on Earth. The first, *mycoplasm*, gave rise to all bacteria and fungi (with the exception of phycomycetes), blue-greens, and organelles of the cell: chromatophores and nuclei. *Amoeboplasm*, so called because it is "namely in the amoeba that all typical features [of this plasm] most clearly and boldly emerge," lies at the foundation of all animals and plants on Earth (Merezhkovsky 1909a, 12–13). However, mycoplasm is also present in the cells of all animals and plants; in the cells of animals it is found in the nucleus, whereas in plant cells it is found in the nucleus and plastids. Six chapters of the book are devoted to examining the differences between the two plasms.

To substantiate his assumption, Merezhkovsky offered extensive data from various branches of biology, displaying his broad knowledge and interpretive ability. A few examples of principal importance for the theory, which enable us to discuss the general character of his proof, are covered here. Thus, proving the profound difference between amoeboplasm and mycoplasm in relation to oxygen, the author cites the works of L. Pasteur, who established that among microorganisms exist anaerobes. The broad distribution of anaerobic bacteria in nature permits one to conclude that bacterial plasm, as part of the mycoplasm, could live without oxygen. This property of mycoplasm has exceptional import, insofar as it permits one to assume that anaerobiosis was a primary condition of the first bacteria—*protobacteria*. Fungi and cyanobacteria, although suited to live in an oxygen environment, are likewise composed of mycoplasm. This condition poses a well-known problem for the theory of two plasms, but this feature is an exception to the general rule, for a whole series of additional features exists, indicating the fundamental difference of the two plasms.

The most important feature of mycoplasm, according to Merezhkovsky, is its ability to synthesize complex organic substances. If this assertion is absolutely clear in relation to animals, then in relation to plants it requires additional explanation. As is well known, chromatophores, which enable plants to build organic substances from inorganic ones, "do not belong to the plant itself" but are "foreign organisms from the kingdom of mycoids, having taken root from without in the protoplasm of the living cell and having entered with it into a symbiosis, called the plant" (ibid., 36). The cytoplasm of the plant cell, just like animal cytoplasm, is incapable of synthesizing organic substances. This is precisely why plants cannot be viewed as typical autotrophs. This part of the argument for the theory of two plasms, it seems, exposes its speculative and contradictory character in an especially clear manner.

Merezhkovsky collected evidence of profound differences between the two large groups of organisms, which he called *mycoids* and *amoeboids*. The fundamental properties of these two are distinct enough to conclude they are completely independent plasms that differ in function and structure. The idea of two worlds of living beings served as the basis for the theory of two plasms. Later, although Merezhkovsky did not work on the theory itself, they had great significance for his systematizing the organic world and understanding evolution. Again and again he returned to its fundamental corollaries.

Recognition of the existence of two plasms carried with it his conviction that the first bearers of these plasms, in various epochs in the Earth's history, emerged independently. The first organisms, consisting of mycoplasm, were bacteria. In the course of time, larger and more complex bacteria were formed from these primary, simple biococci and from their complex descendants—fungi and blue-greens. Amoeboplasm emerged later. Amoeboplasm appeared in the form of small "monera without nuclei" that move like amoebae and feed on bacteria.

The theory of two plasms necessarily carried with it recognition of symbiosis as the cause of the origin and evolution of organisms. Merezhkovsky suggested that, insofar as animals and plants consist of amoeboplasm but include mycoplasm as well, one cannot assume that all living beings arose only through the gradual complexification, divergence, and differentiation of initially homogeneous forms without regard for their possible union and symbiosis. Further, he assumed that at the dawn of life the first mycoid forms were bacteria, which were eaten but not always digested by the amoeboid monera, that is, by the amoeboplasm that lack nuclei. Remaining inside the body of the amoeboplasm, these bacteria sometimes entered into functional relations with them. In time, such symbiotic micrococci formed groups in the center of the amoeboplasm, which became a cell nucleus.

Merezhkovsky viewed the emergence of a primary cell with a formed nucleus as the result of the amoeboplasm that entered into symbiosis with earlier mycoids (bacteria). Amoeboplasm and bacteria, having united and formed in a physiological relation, a single independent organism, gave rise to flagellates and amoebae [protists]—initial forms of the animal kingdom. In the process of evolution, however, still a second stage of symbiosis came into being. An establishment of new representatives of mycoids inside amoebae and flagellates occurred—blue-greens—themselves evolved from the earliest mycoid bacteria. By the time blue-greens entered into coexistence, they had already resulted in a plethora of forms. This second symbiosis gave rise to several separate branches of the plant kingdom that had emerged independently of one other.

Finally, from the theory of two plasms (based on the principle of symbiogenesis, came a new general taxonomy of the organic world, which required reexamination of traditional phylogenetic schemes. Merezhkovsky believed that three kingdoms of organisms existed: mycoids [including his monera without nuclei, bacteria, biococci, amoeboplasm, cyanophyceae, and so forth], plants, and animals. He justified designating a kingdom of mycoids with characteristics of the group: bacteria, fungi, and blue-greens that were *not* the result of symbiosis but had directly developed from the earliest mycoids. The plant and animal kingdoms arose as a result of symbiosis—animals through a single symbiosis [mycoids with amoeboplasm], and plants through double symbiosis [mycoids with amoeboplasm followed by blue-greens].

The theory of two plasms required abolishing the kingdom Protista, which in some phylogenetic schemes was considered a natural group. The protist kingdom was not a natural group, "for there is no middle ground between symbiosis and nonsymbiosis. Either symbiosis with cyanophyceae exists, and then one has plants, or it does not exist, and then we have animals" (Merezhkovsky 1909a, 96). This, in short, is the theory of two plasms.

The two plasm theory was not discussed seriously by Merezhkovsky's contemporaries or others in following years. Most likely it was ignored because it seemed "completely fantastic and unscientific" (Takhtadzhyan 1973, 23). Two of its consequences, primarily questions of taxonomy, were nevertheless at the center of researchers' attention. One concerned the naturalness of the three large kingdoms: mycoid, plant, and animal. Merezhkovsky's conviction about the natural division in the phyla of the organic world of the third kingdom had been established by earlier researchers and served as the basis of a series of modern phyla (Takhtadzhyan 1973). Another consequence concerns the group Protoctista. A series of authors (Moshkovsky 1957; Ivanov 1968; Takhtadzhyan 1973) agree with Merezhkovsky; they are in favor of liquidating this group and its internal division into: Protozoa and Protophyta. Electron microscopic observations corroborate this view. Sh. D. Moshkovsky (1957) believes that data from electron microscopy show that, even at the very lowest level of organization of eukaryotes—the level that does not reveal even the presence of flagella [undulipodia]—organisms of plant and animal nature are readily distinguished. "Already at the given level the basic structural-chemical and physiological differences between plant and animal organisms are clearly exposed: the presence or absence of plastids, the character of exchange processes, peculiarities of change in forms" (Moshkovsky 1957, 133). Alternative opinions exist about the expediency of preserving protists as a single phylogenetic and natural group, from which Metazoa and Metaphyta emerge (Dogel' 1951; Whittaker 1969; Genkel' 1974).

Acknowledging the reality of plant and animal kingdoms, Merezhkovsky believed that such a division was insufficient. In his opinion, more fundamental differences exist among extant organisms—more fundamental than those by which plants and animals are distinguished from one another. These [differences] are concerned mainly with the inner structure of the organism's cells. Thus, bacteria, blue-greens, and fungi—that is, Merezhkovsky's mycoid group—are characterized not only by specific properties of plasm [the absence of intracellular mobility] but also, in contrast to all other organisms, by their lack of nuclei and chloroplasts.[‖]

The cell of the amoeboids, which has a distinct nucleus, separate from the cytoplasm and cell membrane, is constructed according to another principle—that is, of single-celled and multicellular plants and animals. All plants contain plastids, which are also readily delimited from the cytoplasm. Organisms in the group Mycoidei in Merezhkovsky's scheme, then, lack typical cell organization, while those in Amoeboidei do show typical features of cell structure.

Thus, Merezhkovsky clearly raised the question of the existence of two large, independent groups: prokaryotes and eukaryotes. He was first to ponder that most deep, natural groupings of extant organisms should be not into plants and animals but rather into forms distinguished by their type of cell organization. Yet Merezhkovsky did not confer on these groups any taxonomical rank.

The concept of the natural division of organisms into prokaryotes and eukaryotes is now fundamental in constructing macrosystems of living beings. In Takhtadzhyan's systematics, for example, the differences between prokaryotes and eukaryotes, elevated to the rank of "superkingdom," is of foremost significance. "The Aristotelian-Linnaean division of organisms into animals and plants is gradually making way for a more natural division into prokaryote and eukaryote" (Takhtadzhyan 1973, 23).

Further, Merezhkovsky attempted to explain the emergence of intracellular structure in the evolution of the amoeboids. Derived from the theory of two plasms, the cell's nucleus and chromatophores differed from cytoplasm in a variety of ways. The assertion that, in the course of evolution, the amoeboplasm gave rise directly to mycoplasmic organelles was improbable. The only possible mechanism by which the inner cell structure arose, Merezhkovsky maintained, was symbiosis. "Without symbiosis monera without nuclei would have been doomed to remain the same low forms of life, which they were initially before

---

[‖]Merezhkovsky's attempt to bring together and include bacteria, blue-greens, and fungi in one phylogenetic group cannot be recognized as correct. Even the least organized lichens have a series of features that distinguish them from bacteria and cyanobacteria.

bacteria became established in them. We would not have any animal, nor any plant kingdoms with all the infinite diversity of forms; the organic world would have been, on the one hand, a vast and diverse kingdom of fungi, and on the other hand, homogeneous, trifling monera" (Merezhkovsky 1909a, 92–93). Merezhkovsky's idea may be expressed in modern terms: endosymbiosis was the leading factor under whose influence the eukaryotic cell emerged from the primary prokaryotic cell. The concept of the symbiotic origin of plant and animal cells (eukaryotes) was not accepted by Merezhkovsky's contemporaries. Later, it was subjected to severe criticism. Several botanists (Kursanov and Komarnitsky, 1945; Komarnitsky 1947) objected to it. These authors accused the hypothesis of symbiogenesis of being based on analogies; it did not take into account differences in the internal structure of single-celled algae and plant chloroplasts. Plants have ontogenesis, and the resemblance between cell organelles and intracellular symbionts was superficial. The symbiosis hypothesis should be abandoned; the most important achievements in cytology were attained on the assumption of the origin of cell structures by differentiation (Kursanov and Komarnitsky 1945, 254–55). Now these categorical conclusions seem unlikely. The data on cell organelles, which revive the symbiotic hypothesis for the origin of cell structure, were obtained with these same methods of cytology (see chapter 7 for more detail).

A. I. Oparin (1941) criticized the hypothesis of cell symbiogenesis for establishing a sharp boundary between cytoplasm and the cell nucleus, for emphasizing "the preeminent and independent role of the latter," and for denying the genetic integration of cell components. That the "nucleic acids of the nucleus are formed in the end from the cytoplasm," for Oparin spoke against the symbiogenesis theory. Although his criticism is justified for the concept of symbiotic origins of the nucleus, it is unjustified as a criticism of Merezhkovsky's assertion of a sharp boundary between the nucleus and the cytoplasmic organelles. The nuclear genome is exceptionally significant for the preservation and transfer of hereditary information; all properties and features of the organism are controlled by the nuclear DNA, including many steps in the formation of the most important components of plastids (Kirk 1970), which have their own DNA.

The development of chloroplasts from mitochondria—which in the opinion of Breslavets (1959, 1963) was established for the majority of angiosperms—she considered a serious argument against the hypothesis of cell symbiogenesis. The specific features of the morphogenesis of plant plastids and even mitochondria (Nass 1969; Roodyn and Wilkie 1970; and others), where an evolutionary relation between them and bacteria exists and where the possibility of their endosymbiotic origin is seriously discussed, cannot be viewed as weighty as evidence favoring the argument of ontogenetic development of these organelles against the hypothesis of the symbiogenesis of plastids.

Merezhkovsky and Symbiogenesis

The symbiogenesis hypothesis was unfavorably assessed by Koloss (1975). Justly remarking that the transformation of precells into cells is a process not easily explained, he recognized that the essential deficiency of the hypothesis was the absence of an answer about how the "initial nucleus-containing cells" appear. However, Koloss went further, rejecting symbiogenesis as an evolutionary mechanism for the origin of other organelles. Structural and biochemical similarities between cell organelles (for example, mitochondria) and prokaryotes (bacteria), taken "in isolation from the history of development," as Koloss writes, are not sufficient proof of their evolutionary relation. Emphasizing this side of the question, he is silent about the facts concerning the significant functional, biochemical, and genetic autonomy of these cell organelles. The presence of organellar DNA—in contrast to nuclear DNA—Koloss construes to be the result of the divergent development of previously undifferentiated DNA into "two types," which could facilitate a rapid exchange during development, this newly differentiated DNA having received its specific character at various locations in the biont (Koloss 1975, 45). In this debate one may hypothesize other symbiotic interpretations of eukaryotic cell origins. The concept of symbiogenesis has attracted more research attention in the last two decades. The multitude of new data collected in researching chloroplasts, mitochondria, the nature of the mitosis, and kinetosomes of undulipodia using modern electron microscopic, molecular genetic, cytological, and biochemical methods justifies this theory, as Margulis (1970, 11) has justly noted.

Merezhkovsky's entire concept of two plasms, without a doubt, is highly speculative. Although Merezhkovsky was convinced of their diphyletic origins, his assertion of the presence in nature of two independent, living plasms was farfetched. Almost all of the material used as proof of the distinctive features found in bearers of the two plasms could be interpreted differently in the world of modern data. Yet the concept of two plasms contains opinions that clearly deepen and expand our knowledge of biology. A new approach to the formation of taxa was formulated, allowing not only a boundary between prokaryotes and eukaryotes, but also acknowledgement of various evolutionary levels of cell organization. Merezhkovsky attempted to explain the origin of the eukaryotic cell. The possible origin of intracellular structures by the symbiosis of prokaryotic organisms of diverse origins is rational.

Later Works: Features of Symbiogenesis

Symbiosis as the basis of evolution continued to characterize the core of Merezhkovsky's theoretical evolutionary views—whether concerning the symbiogenetic origin of chloroplasts or as the basic principle for the construction of phylogenies.

Merezhkovsky and Symbiogenesis

In 1909 and 1910, Merezhkovsky's books *A Concise Course On Cryptogamic Plants* and *A Concise Course On General Botany* (textbooks for university students) were published. Merezhkovsky's last article, "The Plant as a Symbiotic Complex," appeared in Geneva in 1920.

Assuming symbiosis to be a mechanism of evolution, he tried in his course on cryptogamic plants to establish a phylogeny of the plant kingdom and to solve the question of their origins. The colorless heterotrophic mastigophora he named as the most probable plant ancestor. Ensuing evolution involved blue-greens acquired symbiotically by the colorless mastigotes; algae appeared as a result of such association. The origin of algae seemed to him to be a polyphyletic process: blue-greens, composing a varied collection of pigments, entered into symbioses with colorless mastigotes; and also blue-greens of one type established themselves in mastigophorans that possessed one or two equal [isokont] or two unequal [heterokont] flagella [undulipodia].

A special section in his general botany course was devoted to symbiogenesis theory. Merezhkovsky described how, through deduction, he arrived at the concept of the symbiotic nature of chromatophores (1910, 66). His was a logical conclusion from a series of facts. Presented so categorically, Merezhkovsky's conviction seems groundless. The basis of these ideas was evidence, attained by his research on diatoms. Proof of the symbiotic nature of chromatophores was formulated according to the scheme featured in his 1905 article.

The idea of the polyphyletic origin of algae, attractive as a new argument, distinguishes the 1910 work. The conclusion that algal types originate independently from seven to eight different colorless mastigophora rendered impossible the assumption of the origin of plastids by cytoplasmic differentiation. Accepting symbiogenesis makes highly improbable the specific differentiation of self-sufficient photosynthetic plastids independently seven or eight times in phylogeny (Merezhkovsky 1910).

Until the end of his life, Merezhkovsky remained convinced of the evolutionary significance of symbiogenesis in the origin of cell structure. A lengthy article, "The Plant as a Symbiotic Complex" (Merezhkovsky 1920), documents this fact. In this article, Merezhkovsky first attempted to integrate his opinions with ideas about heredity, beginning with the idea that all hereditary features are regulated by the nucleus. If this is true, he reasons, then the only correct explanation for the continuity of plastids is the one he already suggested in 1905. If plastids evolutionarily emerged by cell differentiation, then this process proceeded gradually. Each stage of plastid formation, like any hereditary feature, should have been consolidated and entered into "the structure of the chromosomes." After becoming genuine cell organelles, plastids should emerge anew with each cell division under the influence of the nucleus. The continuity of plastids, however, indicates the opposite. "To state that they [chromatophores—L. Kh.] once appeared anew, is to state, that they became hereditary, that is, that even today they are emerging anew" (ibid., 28).

Defending the idea of the continuity of chromatophores in the face of the concept of nuclear heredity, Merezhkovsky came close to the concept of the genetic individuality (autonomy) of chloroplasts. He especially compared plastids and mitochondria; such a relationship is important for the symbiogenesis theory, insofar as it will determine the extent of the independence of chromatophores. If the mitochondrial origin of plastids were corroborated, then not only the "theory of the symbiotic nature of the plant cell, but also the entire theory of symbiogenesis will be destroyed" (ibid., 88). The existence of two independent but, in their histochemical properties, similar cell inclusions, Merezhkovsky felt, was a most important proof in support of his theory. He concluded that all assumptions and arguments brought forth in earlier works could not cast doubt on the soundness of the hypothesis of symbiogenesis in the origin of the plant cell.

The role of symbiosis in evolution, an idea he developed over the course of seventeen years (1903–20), stands out in all of Merezhkovsky's work. He suggested the term symbiogenesis in 1909 and later gave it detailed definition: "I called this process symbiogenesis, which means the origin of organisms through the combination and unification of two or many beings entering into symbiosis" (Merezhkovsky 1920, 65). Symbiogenesis as an evolutionary principle permitted one to pose and then solve the question of cell origins and organism evolution. Merezhkovsky formulated the evolutionary concept that he called the theory of symbiogenesis. "So many new facts have arisen from cytology, biochemistry, and physiology, especially of smaller organisms, that an attempt once again to raise the curtain on the mysterious origin of organisms appears desirable. I have decided to undertake such an attempt, and my present work . . . consists in a preliminary exposition of a new theory on the origin of organisms, which, in view of the fact that the phenomenon of symbiosis plays a leading role in it, I propose to name the theory of symbiogenesis" (Merezhkovsky 1909a, 7–8).

Having created the concept of symbiogenesis, Merezhkovsky pondered where it belonged in the general system of knowledge about organic evolution. Above all, he thought it necessary to correlate it with the teaching of Darwin. This he did with the utmost brevity. Darwin's theory, along with the teachings of Haeckel and Nägeli, seemed an unsuccessful attempt to solve this problem. Merezhkovsky considered these theories outmoded because they were based on old data. Newly collected data, then, demanded a new evolutionary conclusion. Theories asserting evolution to be a goal-oriented process, realized on the basis of some imminent beginning, did not contradict his concept but rather motivated it.

Merezhkovsky and Symbiogenesis

Merezhkovsky, in his scientific career, was quite contradictory. Speculation and natural philosophical ideas were not alien to him; his ideas frequently bordered on the purely speculative and even the fantastic. On the general theory of evolution, Merezhkovsky was not original; he agreed with the idealistic doctrine of teleogenesis. He exhibited acumen toward wide, comparative observation, toward general summaries of data both personally acquired and from the literature. He possessed the capacity for the logical grounding of theoretical framework and the rare gift of scientific foresight. Merezhkovsky's ideas were of fundamental significance to the formulation and proof of the hypothesis of symbiogenesis.

4

ELENKIN'S
CRITIQUE OF THE
SYMBIOGENESIS CONCEPT

The most extensive criticism of symbiogenesis in Russian biology was undertaken by an outstanding botanist, Alexandr Alexandrovich Elenkin (1875–1942). He first opposed the concept in 1907. This negative attitude toward symbiogenesis characterized Elenkin's subsequent works, in which questions of the nature and classification of cyanoses, general questions of symbiosis, and lichen evolution were treated. Only toward the end of his life did his attitude change. The notions *lichen* and *lichenous symbiosis* were interpreted in the light of Darwin's work (Elenkin, 1940). A serious attempt was made by Elenkin to reevaluate the entire concept of symbiogenesis.

Elenkin's broad philosophical and biological erudition enabled him to move beyond mere special studies and on to theoretical generalizations (Savich 1944; Komarnitsky and Lipshits 1945; Lipshits 1950; Polyansky 1957; Gollerbach and Sedova, 1974; Trass 1976).

Elenkin's views on the evolution of the organic world were original. They developed as he considered the results of his own experiments on symbiotic relationships in plants, as well as general philosophical studies that interested him.

Treating symbiogenesis, Elenkin leaned on his own interpretation of the relation between lichen components. He proceeded from an evolutionary concept based on the idea of dynamic equilibrium.

The Lichen Thallus and the
Theory of Endoparasitic Saprophytism

The assertion that the fungus-alga association in a lichen is not mutualistic, and therefore [that] the lichen cannot be considered an independent organism, served Elenkin to oppose the hypothesis of symbiogenesis. Having rejected the mutualistic hypothesis of symbiosis, he rejected as a consequence symbiogenesis.

Elenkin opposed the mutualistic theory of symbiosis applied to lichen components at the beginning of the century (1901, 1902). His investigations of many species of lichens served as the basis of this work. Having studied the structure of the thallus, Elenkin discovered a rather thick layer of dying gonidial cells, often

Aleksandr Aleksandrovich Elenkin (1875–1942)

exceeding in number the cells in the living gonidial tissue. This layer of empty cells he called the necrotic zone. A group of lichens with *Pleurococcus* algae (*Pleurococcus vulgaris* and *Cystococcus humicola*) had three necrotic zones: epinecrotic > zoonecrotic > hyponecrotic (Elenkin 1902). Several genera of lichens (*Lecanora, Ochrolechia, Lecidea*) had only one layer, the zoonecrotic, and the genera *Pertusaria* and *Variolaria* had all three layers. New observations of the dying gonidium in the species *Biatora armeniaca* and *B. aenea* were described (Elenkin 1904). Both species had well-developed epinecrotic and hyponecrotic zones, including a thallus layer in which the number of dying gonidia cells exceeded the living. By counting, the correlation was established to be a ratio of three to two (dying:living cells) in the bionecrotic layer (Elenkin 1904).

The presence of vigorous necrotic zones in the lichen thallus and the distribution of the phenomenon in heteromerous lichens attested to Elenkin the unequal relations between lichen components. The theory of mutualism was groundless; the data demanded another explanation of the nature of the interrelation. Elenkin originally believed that the relation could be described as the *autotrophic* endosaprophyticism of the fungus on the alga (1901, 1902). The fungal component feeds at the expense of the organic matter in the dying and decomposing cells of the lichen gonidium. Later studies, however, showed penetration of the fungal haustoria into the gonidial cells, demonstrating a parasitic method of fungal nutrition [necrotrophic] at the expense of the alga side by side with saprophyticism [symbiotrophy]. Haustoria in the gonidia show, wrote Elenkin (1904, 39), that "endosaprophyticism contains within itself an element of parasitism." The penetration of the fungal haustoria into the gonidia was later confirmed by A. N. Danilov, who revealed haustoria in various stages of development in the gonidia of many crustose, foliose, and fruticose lichens. He showed that the hyphal net in the gonidia were continuous with the receptive fungal hyphae of the lichen. Elenkin regarded these results highly (1910).

He (1906) generalized his morphological results, developing the theory of endoparasitic saprophytism. This theory provided the most acceptable explanation of lichen phenomena, interpreting the interrelations of the lichen components as antagonistic.

This interpretation was a vital contribution to the theory of lichen symbiosis (Polyansky 1957; Gollerbach and Sedova 1974). But Elenkin went further, deducing a lack of unity of the lichen components. He began to deny the integrity of a lichen as an independent organism.

The clearest progress of his argument is Elenkin's 1907 article "The Relation of Lichenous Symbiosis to the Evolution of Plants." From the theory of endoparasitic saprophytism, lichen symbiosis represents a typical parasitic association. No parasitic symbiosis, no matter how integrated, has ever been considered to be on a path leading to the formation of morphological unity as an organism or formation of a separate taxon. There are not, therefore, nor can there be, any grounds to assert that the lichen is an independent, complexly structured, organism.

Elenkin surveyed Famintsyn's arguments from this point of view to support his concepts. Most important, in Famintsyn's opinion, was the existence of a complex, but integral, organism of the lichen, arising historically from two distinct forms. Elenkin confirmed that, in any parasitic association, two separate organisms exist—the parasite and the host. Famintsyn stressed the existence of specific traits in lichens that were the results of the joint activity of its constituents—for example, soredia, the special organs of reproduction; Elenkin responded that the formation of soredia is only a minor change from the characteristics of any parasitic symbiosis. This trait, the dependent "peculiar position of the host within the parasite," in his opinion, represented one case of "the complete disorganization of the fungal component and the liberation of the colony of alga" (Elenkin 1907a, 169). The other cases of symbiosis that Famintsyn introduced— for example, the appearance of endomycorrhizae—Elenkin regarded as typical "unstable" parasitic associations and therefore not supportive of the hypothesis of the evolution of new independent organisms by symbiosis. Elenkin concluded that lichens and other symbioses "can on no account serve as factual proof of the theory of evolution" by symbiosis (ibid., 170). Elenkin denied the entire concept for many years. His negation was this: the type of relation between lichen components is the main criterion of its integrity, and therefore the presence of parasitic, antagonistic relations inside the lichen symbiosis precludes viewing the lichen as an integrated individual, analogous to other plant life.

In contemporary lichen literature, the question of the relation between fungus and alga in the thallus remains open, but even at that time the existence of a complex, lichenous, individual organism was irrefutable.

The interaction of the components in a lichen symbiosis is not now seen as constant and uniform for all species. All types of relations occur—from the most casual associations with hardly noticeable (and, at the earliest stages of development, still facultative) parasitic relations, to firmly secured, obligate symbiosis (Oksner 1956). Several hypotheses attempt to explain the double individual found in lichens. Nonetheless, it is still not possible to specify which description—balanced or destructive parasitism, helotism [see p. 81], biotrophism, mutualism, commensalism, consortism, endoparasitic saprophytism, phycosaprophytism, or other—is definitive (Trass 1973).

One must face the complexity and heterogeneous character of the relation between the lichen components. Must this lead to the rejection of the concept of individuality for lichens? Many facts undoubtedly attest to the highly organized integrity of some lichen symbioses. Lichens have "two organisms with contradictory abilities . . . living in unity, interpenetrating one another and, finding themselves in constant struggle, forming a new complex organism, developing on the basis of this struggle of opposites, but having new, specific capabilities and new patterns of development" (Kursanov and Komarnitsky 1945, 436). The individuality of a lichen is expressed in the existence of unique structural and functional traits that are absent in the constituent components. The high level of organized integrity in lichens is seen not only in their bodies but also, and more importantly, in tracing their evolution.

The individuality of a lichen as an independent living entity is particularly clear for the great specialization of its fungal and algal components. The integration of the bionts resulting from their evolution is so great that it is impossible for them to exist independently. The fungus is a highly specialized, obligate component, but also lichen algae cannot live outside the lichen thallus. The alga *Trebouxia,* found as the component of nearly half of all known lichen associations, does not exist independently (Ahmadjian 1971). This fact is explained by one of two suppositions: [either] *Trebouxia* existed in the past as a normal, free-living alga and at present survives only in lichen association, or this genus is the result of the profound lichenization of several independent extant algae (Ahmadjian 1971). The first supposition seems unlikely to me (L. Kh.), since efforts to find an independent *Trebouxia* have not met with success. Comparative investigations of the type of chloroplast, the presence or absence of the wall of the zoospore, the mode of nutrition (heterotrophy or photoautotrophy) of *Trebouxia* and representatives of the genus *Chlorococcum*—taking into account all of the characteristics of these algae—allow one to speak, in Ahmadjian's opinion, of a phylogenetic link between *Trebouxia* and free-living algae of the genus *Chlorococcum.*

The existence of complex, parasitic relations in lichen components is not an argument against the possibility of the symbiotic formation of new organisms. Elenkin's critique of symbiogenesis from the theory of endoparasitic saprophytism in lichen symbioses does not appear sufficiently convincing.

Criticism of Symbiogenesis and the Law of Dynamic Equilibrium

In evaluating symbiogenesis, Elenkin proceeded from his own evolutionary idea and formulated the *law of dynamic equilibrium* in 1921. This idea, with the theory of *equivalentogenesis,* elaborated somewhat later (Elenkin 1926), formed the basis of the evolutionary views he held throughout most of his career.

In his world view, Elenkin was a confirmed, and it would not be an exaggeration to say, militant, materialist (Polyansky 1957). He was a mechanist. Elenkin saw the possibility of reducing "all phenomena of nature to physiochemical processes and, in the end, to movement—that is, mechanics" (Elenkin 1921b, 76). The idea of dynamic equilibrium, particularly important to him, enabled him, "without resorting to vitalism" (as he wrote), to solve all problems of the phenomena of nature, including the causes of evolution.

Considering his judgments a particular case of the philosophy of Herbert Spencer, Elenkin thought that evolution consisted of transformations directed at equilibrium with the changing external conditions. The idea of dynamic equilibrium was based on Spencer's hypothesis of development as a constant adaptation of internal relations to external conditions, provoked by the activity of external causes and proceeding by means of direct equilibration (through changes of function in ontogenesis and the formative action of factors of the environment).

The idea of dynamic, or unstable, equilibrium reflects the fact that two or more organisms with various or even contradictory abilities find themselves in constant struggle and form a lasting unit, showing that "some organisms are stronger than others, but they cannot finally destroy the weaker because of the constant fluctuation of external conditions, favoring now one and now the other" (Elenkin 1921b, 43).

Elenkin (1921) created a theoretical conception, formulated as a law, based on the idea of dynamic equilibrium. This law "is defined as a concrete expression of the interaction of antagonistic forces in any living system, equilibrating one another at any given moment" (Elenkin 1921b, 101). The law led to a hypothesis about the direct or functional dependence between the components of an association relative to the conditions of the external environment (ibid., 101–2). The idea was introduced to explain interactions of the components of lichen symbioses. Elenkin, however, later expanded it, viewing it as a driving force in evolution. From the position of the law of dynamic equilibrium, Elenkin criticized Darwinian causes of evolution.

In evaluating Elenkin's evolutionary concept, we see that he never explained why organisms, when faced with changes in the environment, always react adequately and suitably with respect to specific conditions. Elenkin accepts expedient reaction as a self-explanatory phenomenon.

As many biologists have already noted (Polyansky 1947; Shmal'gauzen 1969; Blyakher 1971; Zavadsky 1973; and others), the recognition that organisms directly adapt implies an unstated assumption that this ability is primordial. The theory of dynamic equilibrium, as one version of mechano-Lamarckism, teleologically resolved the problem of the origin of organic expedience. The principal error of the theory of dynamic equilibrium in evolution comes from the fact that it reduces historical regularity to philosophical and even mechanical regularity.

The law of dynamic equilibrium was criticized in Soviet botanical literature (Komarov 1940; Komarnitsky and Lipshits 1945; Komarnitsky 1947; Trass 1973; and others).

Elenkin (1921) critically evaluated the concept of symbiogenesis from the position of the law of dynamic equilibrium; he had two principal objections. Symbiogenesis was unacceptable because, although indirectly it acknowledged the principle of stable or nondynamic equilibrium, its supporters accepted the hypothesis of mutualistic symbiosis. They suggested mutually beneficial interactions, seeing in them "something permanent, leading to mutual advantage" (Elenkin 1921b, 81). Because the mutualistic idea was accepted, symbiogenesis, as represented by Elenkin, was built on a static principle and therefore failed to withstand criticism. This objection, not directly related, did not touch the essence of symbiogenesis. The criticism was directed against an idea accepted by supporters of symbiogenesis only in passing and only in connection with the problem of the nature of the interaction between partners in a symbiosis.

A critique of Elenkin's objection to the theory of symbiogenesis was written in 1928 by V. N. Beklemishev (1970a). In trying to refute the idea of mutualistic symbiosis, Elenkin committed a serious logical error by absolutizing mutualism as a form of bond and then applying it indiscriminately to the phenomena of association. The relation between organisms in symbiosis is exceptionally diverse and complex. The symbiotic bond involves numerous interactions; one speaks only of an extent of parasitism or mutualism in symbiosis. "One must not forget that mutualism has degrees. Also, opinions about the anthropomorphism of the concepts of *mutualism, parasitism,* and others in comparison with the objectivity of the concept of *dynamic equilibrium* are extremely naive" (Beklemishev 1970a, 32).

Elenkin's second objection concerned the factual basis of symbiogenesis. He felt it was built on a mistaken interpretation of the facts because it viewed lichens and other symbioses as tight, obligate associations completely homologous to independent organisms. An individual lichen, Elenkin thought, cannot be considered an independent organism because the factors guaranteeing its life and development of the symbiotic relations are distinguished from those of real organisms. "If the life of a lichenous system submits to the law of dynamic equilibrium and is exclusively regulated by external factors, then the life of the organism, to a large extent, is regulated by 'internal laws'" (Elenkin 1921b, 88). The essential feature of an organism is its great integrity and stability in comparison with a symbiosis (ibid., 88–89). The evolution of an integrated symbiosis and of an organism must be ascribed to phenomena of completely different orders: "Between the evolutionary development of organisms and of symbiosis one may place only analogy, not homology. . . . Although it is possible to create an analogy between these elements, since in both cases we have a synthesis of form, these unities have essentially different origins and different procedures" (ibid., 89). Like the first, this objection cannot be considered just. Drawing conclusions from observations of differences in the life of lichens and typical free-living organisms, Elenkin thought it possible to characterize their phylogenetic differences. This concept means the recognition of the identity of the laws of ontogeny and phylogeny, which characterizes many teleological hypotheses.

The idea of dynamic equilibrium as the theoretical basis of lichen symbioses was criticized by many Soviet researchers. They noted that although its distinct, initial assumptions were indeed correct (for example, the hypothesis regarding the antagonistic relations between fungus and alga), the theory of dynamic equilibrium unilaterally exaggerated the influence of external factors and underestimated "the laws of development and the hereditary peculiarities secured by the protracted phylogeny of lichens" (Komarnitsky and Lipshits 1945, 125).

The principle of dynamic equilibrium, wrote the great Russian geobotanist H. F. Komarov, rested on the negation of the principle of self-development. "A critical view of the theory of dynamic equilibrium does not signify a rejection of the condition of dynamic equilibrium. Equilibrium, however, must be understood not only as relative but also as self-perpetuating" (Komarov 1940, 300).

Elenkin criticized symbiogenesis in other works—for example, in "New Works in Foreign and Russian Journals relating to My Theory of Endoparasitic Saprophytism and the Law of Dynamic Equilibrium in the Components of Lichenous Symbiosis" (1922d). Here a different motif of criticism sounded. Previously, his objections had involved the interpretation of the relations between the components of an association; in this article symbiogenesis is doubted from the view of accepted lichen classification. "I understand and recognize the logical permissibility of the theory of symbiogenesis, but for a number of reasons I cannot acknowledge it as correct" (Elenkin 1922d, 4). What were these reasons? The classification of units of lichens proceeds on the basis of the traits (the structure of the body) of only one of the components—the fungus. "When I say *lichen*, then in a systematic sense, I mean only one organism, that is, a fungus. . . . The aggregate of these epiphytic fungi, related by us to various species, and composing a unique group in mycology, is known under the name *lichens*," concluded Elenkin (1921b, 85–86). The recognition of lichens as integrated organisms led to serious misunderstandings of classification. Even now, however, the problem of lichen classification is not resolved. Serious, multifaceted studies of the nature of these organisms, conducted in recent years, more sharply posed the question of whether their classification takes into account the qualitative, specific traits of the lichen as units and not just the traits peculiar to the fungal components (Trass 1973). Elenkin's assertion that symbiogenesis provides a contradictory basis for the taxonomy of lichens is therefore not sufficiently convincing.

Elenkin's Attitude toward Symbiogenesis in His Last Years

Elenkin's views on the evolutionary process underwent a crisis at the end of the 1930s. I consider here his works from the second half of the 1930s (1936, 1939) as well as [his] large work from 1940, "The Understanding of *Lichens* and *Lichen Symbiosis*," published posthumously in 1975.

Elenkin's conversion to Darwinism was closely related to his changing views on the lichen symbiosis. During much of his life he had been an advocate of the bioanalytical tendency in lichenology. The idea of a lichen as a consortium—in which a fungus and an alga constituted an indivisible, whole, independent organism, corresponding to any green plant—was alien to him. This was connected primarily to his mistaken understanding of the ideas of *mutualism* and *consortium,* which really have little in common (Elenkin 1975, 14). Moreover, the bioanalytical tendency to recognize lichens essentially as fungi, acting parasitically or saprophytically on algae, having identified antagonistic relations between the lichen components, presupposed the possibility of interpreting lichen evolution under the influence of the direct action of the surrounding environment on the basis of the law of dynamic equilibrium. This was expressed in a mechanistic idea of development in which dynamic equilibrium was taken as a universal law explaining all manifestations of life in the organic world. "The aims of the bioanalytical tendency, although based on the factual basis of the theory of the dualistic origin of lichens . . . soon began to be interpreted in an anti-Darwinian sense," explained Elenkin (1975, 20).

Lichen evolution, as understood by the advocates of the bioanalytical tendency, was equated with variability and explained by Elenkin himself from the position of the study of dynamic equilibrium. His subsequent understanding of lichen evolution stemmed from an acknowledgment of Darwinism and the treatment of lichens as integral, living units (ibid., 20). He reasoned thus: Lichen life is regulated by a unity of internal organizations, in which the alga, in the form of the so-called gonidial zone, is the photosynthetic apparatus. The perfection of the internal regulation of lichen components is manifested in the presence of certain specific traits (the ability of vegetative reproduction, the independence of feeding on the substrate, and so forth). These traits guarantee the lichen a great deal of endurance and a place as pioneers among plants with regard to the effective use of the surroundings under extreme circumstances. Only "having laid at the basis of our understanding of lichens the idea of them as unique, integral organisms," wrote Elenkin, "do we get a clear picture of their evolution on the basis of the creative role of natural selection" (ibid., 45).

Elenkin considered lichen evolution based on natural selection in which the gradual deepening of the level of integration, the extension of the connections between the algal and fungal components, and the changing character of the relations between them occurred. Symbiotic relations touch all fungi and all algae found in communities, "but from this mass only a few species of fungi and even fewer species of algae enter into somewhat extended, close contact with one another. The extensiveness of this contact is defined primarily by the degree of that mutual benefit that draws the symbionts into close association with one another" (ibid., 34).

The relations between the components at the highest stages of lichen evolution are defined as endoparasitic-saprophytic. "Protracted endoparasitic saprophytism is the beneficial adaptation produced in the process of natural selection at the first stages of the evolution of symbiotic relations between a fungus and an alga, which also created the possibility of the further phyletic evolution of lichens as complex and not simple symbiotic units," wrote Elenkin (1975, 37). The extensive endoparasitic saprophytism precisely creates the uniformity of the lichen— its uniqueness as an organism in a biological and a taxonomic sense.

Having acknowledged the lichen as a unique, wholly independent, integral organism, Elenkin reconsidered his attitude toward symbiogenesis, agreeing with it in respect to lichen evolution. But, as can be seen from his fundamental 1936 work "Blue-Green Algae," Elenkin continued to maintain a critical attitude toward symbiogenesis as a universal theory. He no longer agreed with the idea of the cell as "a symbiotic complex." He was particularly interested in the problem of the symbiotic origins of chloroplasts. In Elenkin's opinion, the comparative analysis of the blue-green algal endosymbioses (cyanelles, according to A. Pascher's terminology) with plants would be especially interesting. The factual basis of the hypothesis of the origin of chloroplasts by symbiosis depended on how much it revealed, in a long series of endocyanoses, the transitions from free-living blue-greens through cyanelles (blue-greens endobiotic in the plasma of the cells of colorless organisms) to forms in all respects homologous to the chloroplasts of plants. Endocyanelles with plant cells losing their photosynthetic function deserve particular attention. Investigation of such symbioses would allow one to explain the path by which associated blue-greens merged with their host cells, taking on the function of plastids. To the group of the most important endocyanelles with chlorophylless plants, Elenkin assigned the association of single-celled, colorless alga *Geosiphon pyriforme* (with *Nostoc symbioticum* as endobiont), cyanelles of the species *Glaucocystis nostochinearum* and *Gloeochaete,* and those of *Cyanophora paradoxa* [Kies and Kremer 1990 (see p. xxiv here)].

Elenkin and Symbiogenesis

To demonstrate the symbiotic origins of chloroplasts presupposes an explana-tion of the relations between the cyanelles and the host cytoplasm from a morphological, physiological, and biochemical point of view. The biology of the 1930s, however, did not yet have direct experimental data, and the acquisition of these data, a matter for future science, was not a basis for considering cyanelles—even in the most integrated stages of their development—to be homologous to plastids (Elenkin 1936, 178).

Elenkin noted several separate cases in which the symbiogenesis hypothesis might be of scientific interest. "I think it is not superfluous to emphasize specifi-cally," he wrote, "that I do not have anything against this type of opinion (referring to the opinions of Kozo-Polyansky on the role of intracellular sym-biosis in the origin of the organelles of cells—L.Kh.) in a hypothetical form, which, if they are constructed only on factual data, have significance for further investigation, but I do protest against the excessive dogmatism of symbiogenesis" (Elenkin 1975, 11).

With the switch to the position of Darwinism and the change in his under-standing of the nature of lichens, Elenkin's attitude toward symbiogenesis also changed.

We have surveyed here the works of Elenkin in which his critical attitude toward symbiogenesis is reflected. In the first decade, his criticism was from the mechanistic position of dynamic equilibrium; therefore it cannot be acknowl-edged as correct. In the last years of his work, his evaluation of the scientific value of the symbiogenesis hypothesis became more objective.

In conclusion, although Elenkin was the most serious opponent of the sym-biogenesis concept, his research on lichens had great, primary significance for its further elaboration. First, he constructed a phylogenetic series of gelatinous lichens, which came to be considered one of the direct proofs of the reality of the process of symbiogenesis (see chapter 1). His analysis of blue-greens proposed a model for the study of the evolution of chloroplasts that has basic significance for symbiogenesis hypothesis.

Finally, Elenkin formulated in a precise manner the idea of the transition of the relations of symbiotic organisms to the physiological relations of a single com-plex organism, also based on the example of gelatinous lichens.

This overview shows that Elenkin's work should be considered an important landmark in the development of the concept of symbiogenesis.

K<u>ozo-polyansky's</u> 5

CONTRIBUTION TO

SYMBIOGENESIS

The work of Boris Mikhaylovich Kozo-Polyansky (1890–
1957) was of major significance for the concept of symbiogenesis during the
1920s and 1930s. A fine botanist-theoretician, he elaborated in great depth the
scientific bases of evolutionary morphology and phylogenetic systematics of
plants (Takhtadzhyan 1950; Kamyshev 1957; Blagoveshchensky 1957; Bazilevskaya 1959; Rutsky 1959; Cheremisinov 1959, 1962; Levina 1965).

Working to solve questions of systematics and plant morphology, Kozo-
Polyansky refined and specified general laws of evolution. *Fundamental Biogenetic Law from a Botanical Point of View* (1937), *The Problem of Mimicry in
Botany* (1939), and *Laws of Plant Phylogenesis and Darwinism* (1940) were his
most important works.

Kozo-Polyansky's scientific career included works, which began in the 1920s
(Kozo-Polyansky 1923, 1925a, 1925b, 1926, 1932, and others), specifically
devoted to evolutionary theory. R. E. Levina (1965) believes that Kozo-
Polyansky's scientific career had a single direction: the verification and development of Darwin's evolutionary teachings.

If many facets of Kozo-Polyansky's scientific career have already been documented in more or less detail, a single most important aspect of his evolutionary
ideas, [including] those on the evolutionary significance of symbiogenesis, was
overlooked and is practically unexamined in contemporary literature. Yet symbiogenesis is nearly central to Kozo-Polyansky's general theoretical outlook.
Over many years he collected data from the literature to serve as a basis for
symbiogenesis. He developed theoretical approaches to analyzing problems and
strove to connect the theory of natural selection and the hypothesis of symbiogenesis. His recognition of the evolutionary significance of symbiosis in many
respects determined the understanding of basic evolutionary factors. Kozo-
Polyansky approached the problem of individuality in plants and the biogenetic
law, as applied to plants, from symbiogenesis.

The lack of criticism of Kozo-Polyansky's ideas can be explained by the fact
that the symbiogenesis hypothesis itself was considered unscientific fantasy
(Takhtadzhyan 1973).

Boris Mikhaylovich Kozo-Polyansky (1890–1957)

Details of His Work on the Concept of Symbiogenesis

Kozo-Polyansky worked directly on symbiogenesis over a comparatively short period, the first half of the 1920s, when his major article ("Symbiogenesis in the Evolution of the Plant World," 1921) and his monograph (*A New Principle of Biology: An Essay on the Theory of Symbiogenesis,* 1924) appeared. His outlook on the evolutionary role of symbiosis assumed its final form in these works; Kozo-Polyansky's views remained essentially unchanged in subsequent years. In later works on evolutionary problems and specific botanical questions, he merely refined and supplemented his earlier opinions with new data and responded to criticism.

Kozo-Polyansky first stated his views on symbiogenesis in a lecture at the opening of the First All-Russian Conference of Botanists in 1921. Here he emphasized that accepted ideas about the phylogenetic links among several groups of organisms required reexamination from the viewpoint of symbiogenesis. The emergence of new taxa does not always occur by means of differentiation of the original forms. This process not infrequently requires the synthesis of the preexisting taxa (for example, in hybridization and symbiosis). Therefore, making a judgment about the phylum as a "family tree" as completely iso-morphous with the topology of a real tree would be false. The graphic representation of the plant phylogenetic system is not identical to the topology of a real tree and differs in "the converging and merging of lines" [anastomosis] (Kozo-Polyansky 1921a, 101). Kozo-Polyansky indicated that, in many cases, searches for intermediate forms are unnecessary: according to symbiogenesis, the emergence of new taxa occurs not by the accumulation of minute hereditary changes but by a single transition from original to offspring. One reason for Kozo-Polyansky's interest in symbiogenesis was an aspiration to define phylogenetic ideas more precisely, reexamining the principles of the construction of plant phylogenies. The necessity of solving particular phylogenetic problems directed his thought to the problem.

One must, however, seek deeper reasons for Kozo-Polyansky's interest in symbiogenesis in the historical circumstances of evolution studies in the 1920s. This period was characterized by the fact that, although a number of serious impediments that had faced Darwinism had been eliminated, still "the study of evolution by means of natural selection in fact remained at the stage of probable hypothesis" (Zavadsky 1973, 319). The greatest discoveries in genetics made in this century (the rediscovery of the laws of Mendel, the chromosomal theory of heredity, and the study of mutation) actually strengthened the position of Darwinism. New factual material, such as the elucidation of the adaptational significance of the protective coloring in insects (the work of A. Chesnola, 1904, C. Swinnerton, 1916, and others) or the well-known experimental data collected in 1909 and 1913 by N. V. Tsinger, supported Darwinism by favoring the theory of natural selection. Nevertheless, we cannot prove that this new data had greater weight than the varied evidence used to form the basis of anti-Darwinist conclusions (Zavadsky 1973). Toward the end of the 1920s, Darwinism experienced a crisis of experimental proof. The possibility of directly examining natural selection still did not exist: the mutational process and the laws of gene flow and their complexes in natural populations were not yet studied. Evolutionary ecology in the contemporary sense did not yet exist. All of this emphasized by the 1920s the well-known weakness or, better stated, the inadequacy of Darwinism.

Kozo-Polyansky perceived a need to obtain direct proof of the process of evolution by natural selection. His first work, a 1922 article entitled "The Finale of Evolution," was devoted to problems of evolution theory. Direct, experimental proofs of evolution, he pointed out, are still insufficient, although their significance for justifying evolutionary ideas is enormous and even decisive. Kozo-Polyansky turned to symbiogenesis as direct proof of the evolutionary process and as a means of eliminating the same impediments that hindered Darwinism. In his opinion, the easily verified fact of the emergence of lichens, which are complexly structured and well-adapted plants, through natural synthesis (symbiosis) from relatively simpler and less adapted forms, served well to substantiate Darwinism. Symbiogenesis elevated evolution from probable hypothesis to experimentally substantiated theory. "Until now the generally accepted belief persisted," he wrote, "that the process of evolution could be substantiated only in an indirect manner, that this process did not allow for direct and even experimental proof. . . . But such examples as the emergence of lichens from the combination of algae and fungi do nothing if not enable this sort of direct, experimental examination, illustrating the emergence of more complex and better adapted organisms from relatively simple and less adapted ones" (Kozo-Polyansky 1924, 127).

In his first article devoted to symbiogenesis, Kozo-Polyansky set two tasks for himself: to clarify the relationship between symbiogenesis and Darwin's ideas and to demonstrate the role of symbiosis in the plant world and its evolution.

Accepting Darwinism as the sole scientific theory of evolution, Kozo-Polyansky (1921b, 3) asserted that symbiogenesis, as developed by him, not only did not contradict Darwinism but actually supported it in explaining the driving forces of evolution. Based on natural selection, symbiogenesis opens a new approach to understanding the mechanism of evolutionary transformation in the living world. Symbiogenesis promoted evolution by the unification of several original forms into a single, complex system functioning as one whole.

The great prevalence of varied forms of association illustrates the universal significance of symbiosis for the plant world and its evolution. Kozo-Polyansky viewed the plant cell as a symbiotic complex and concluded that all chlorophyll-bearing plants, the "entire green kingdom, is a gigantic example of symbiogenesis" (ibid, 14). The universality of symbiosis was verified by its geological antiquity and its biogeographical significance (ibid., 20–21). Kozo-Polyansky recognized this universality in a wide sphere of activity. Symbiogenesis functions successfully on cell, population, species, and community levels of the organization of life.

The question arises of how to reconcile Kozo-Polyansky's views on the universality of symbiogenesis with natural selection as the driving force of evolution. Are his views inconsistent with or even a retreat from Darwinism?

By asserting the universal significance of symbiosis, Kozo-Polyansky strongly objected to recognizing it as the driving force of adaptive evolution. The universality of symbiosis in evolution did not mean, he persistently emphasized, an interpretation of symbiosis as the universal mechanism of evolution, equal to the effect of natural selection. Nevertheless, Kozo-Polyansky's "universalization" of symbiogenesis gave rise to sharp criticism and a negative evaluation of symbiosis research (Genkel' and Yuzhakova 1936; Kursanov and Komarnitsky 1945; Genkel' 1946; Komarnitsky 1947; Bazilevskaya 1959, and others). Undoubtedly, Kozo-Polyansky's judgment of the universal significance of symbiosis is an extreme exaggeration, but at the same time I suggest that evaluation of the study of symbiogenesis requires distinction of the concept of symbiogenesis as important for evolutionary theory and the erroneous claim that symbiosis is the universal principle of evolutionary change.

Kozo-Polyansky returned again and again in later works to the question of evolution and symbiogenesis. He defended symbiogenesis and in 1925 evaluated Famintsyn's work in relation to it. He viewed it as the confirmation of the possibility of "a direct and experimental determination of the reality of evolution." In "Introduction to Darwinism" (1932), Kozo-Polyansky tried to explain why most geneticists, including E. Baur, pass over symbiogenesis in silence. The only explanation from his viewpoint was that symbiogenesis was difficult to confine within the boundaries of prevalent ideas about evolutionary progress.

In *The Fundamental Biogenetic Law from the Point of View of Botany,* Kozo-Polyansky examined individuality in plants and distinguished six fundamental levels of individuality; the first he designated plastid and cell. He considered the widespread distribution of intracellular microbial symbioses to be the most important fact in support of the argument for the individuality of organelles. An exceptional similarity exists between several cell organelles and free-living microorganisms, along with great autonomy and genetic continuity of plastids; finally, some microorganisms demonstrate a propensity to form unions serving as good "models" of cells (Kozo-Polyansky 1937, 197). Even though the hypothesis of the symbiotic origin of plastids is doubted because plastids are not capable of independent existence, Kozo-Polyansky emphasized, it must not be expected that an organism composing part of a cell will preserve qualities of its free-living antecedents; indeed, over time, the capacity to live independently could be lost. He was particularly attentive to symbioses of algae with sponges and coelenterates—that is, to symbioses in the lowest [sic] phyla of the animal world. He suggested that animals with their false "chlorophyllic nuclei" could represent one of the first attempts of nature to construct the type of organism now typical of the plant world (ibid., 199). Here Kozo-Polyansky responded to the objection that understanding cells as systems that have emerged as a result of the integration of individuals of lower rank is a mechanistic understanding of the evolutionary process. Such, in part, was the opinion of G. K. Meyster (1934). This objection cannot be justified in any measure, wrote Kozo-Polyansky, since the symbiotic association of organisms in one living system is not a simple "sum" of the components; indeed, as a result of symbiogenesis, "in all cases a qualitatively new whole emerges, and the components cease to be themselves" (Kozo-Polyansky 1937, 203). In addition, the conviction that the cell emerged historically through the combination of the most simple representatives of the organic world, does not exclude the principle of differentiation of the system and its components, which proceeds in connection with this integration.

A Definition of Basic Concepts

Kozo-Polyansky devoted much attention to describing the basic ideas of symbiogenesis. He rightly felt that defining them more precisely would clarify the problem. There were three such ideas: symbiosis, consortium, and symbiogenesis.

Kozo-Polyansky proceeded in his definition of symbiosis from the fact that organisms entering into close joint existence derive great biological benefit and ecological advantage from the association. This advantage expresses itself in the enhanced survivability of the symbionts and in the biological prosperity of symbiotic groups (Kozo-Polyansky, 1921b). Symbiosis is a special form of adaptation, ensuring the biological progress of organisms—that is, of its participants. Thus, by discussing the concept of symbiosis in the Darwinist sense, Kozo-Polyansky arrived at a conclusion about the selective value of symbiotic combinations.

The specific nature of symbiosis as a phenomenon wherein two or more organisms coexist in a single system required a more precise definition, taking into consideration the mutual relations between partners. Kozo-Polyansky considered that, as a rule, the principle of mutual benefit (mutualism) lies at the foundation of symbiosis. In this case, both partners derive benefit from joint existence. But symbiosis could also be built on the basis of one-sided benefit (commensalism). Kozo-Polyansky demanded that mutualism be clearly delineated from the anthropomorphic concept of "friendly" symbiosis because the latter, in his opinion, was the result of a transferral from the human world to the world of animals and plants. "Mutualism has nothing in common with altruism," he wrote, "the latter suggests a voluntary and conscious mutual aid; mutualism comes into effect automatically and mechanically, unconsciously on the basis of the 'chance' correspondence of interests" (Kozo-Polyansky 1924, 18). If symbiosis gives advantage to only one symbiont, then this benefit could be based not only on the exploitation of another organism as a place of residence (domatia and synoikia) or for defense against enemies, temperature, and so on, but also as a source of nutrition (parasitism). If one proceeds to analyze the phenomenon from the viewpoint of the reciprocity of the organisms, one must understand the term *symbiosis* to encompass all forms of coexistence, beginning with mutualism and ending with typical parasitism.

Kozo-Polyansky designates one feature of symbiosis to express most precisely the essence of the given phenomenon—the level of integration of the components in a symbiotic system. The level of integration is expressed in the appearance of new morphological features and physiological properties specific to the symbiotic system and absent in its constituents; this level of connection depends on how the synthesis of organisms occurs. Thus, in the case of simple *apposition,* the extent of integration may be negligible—the symbionts are located nearby (for example, in the case of intercellular symbiosis); the method of *inculcation* emerges as a closer tie—one symbiont penetrates into the cells of the other (intracellular symbiosis); and the highest level of unity is achieved with merging symbiont cells (for example, in the sexual process). Coexistence at any level of connection among organisms, according to Kozo-Polyansky, is known to be advantageous in the struggle for existence. The level of integration, which has become an adaptationally valuable feature, can steadily increase—that is, selection favors its advancement.

To understand the evolutionary role of symbiosis, Kozo-Polyansky suggests using the concept of consortium—the combination or system of more simple and homogeneous organisms, which, while preserving in large measure the individuality of its components, presents itself morphologically, physiologically, and ecologically as a single organism (Kozo-Polyansky 1921b, 6; 1932, 60). Consortia completely correspond in content to *formative symbioses,* a concept introduced by Famintsyn (see chapter 2). In contemporary community ecology, the concept of consortium designates a special unit of the community: the combination of a population of a specific type of autotroph and the aggregate of organisms involved in autotrophic, mixotrophic, and other relations (Rabotnov 1969; Yaroshenko 1975). Thus, in contrast to the way that community ecologists treat the concept of consortium, Kozo-Polyansky raises the concept of consortium to the organism level and designates symbiotic forms of life as such.

According to Kozo-Polyansky, the consortium is qualitatively a new system and not a simple sum of properties of those organisms that constitute it. It is "vitally more stable and more complete" than each of its separate components. A consortium is constructed on the mutually beneficial relations of its components. The persistence of consortium members from generation to generation is ensured thanks to the existence of special reproductive organs and by means of infiltration by the guest symbiont into the presumptive sexual cell rudiment of the host symbiont. In some cases persistence is assured by a type of cytoplasmic inheritance (Kozo-Polyansky 1932, 42).

The emergence of consortium members (symbiotic organisms) is a protracted historical process, adjusted at each of its stages. This phylogenetic emergence of new organisms can be called *symbiogenesis*. Unveiling this concept in more detail, Kozo-Polyansky (1925b) wrote that, by symbiogenesis, he understands the evolutionary process that proceeds by natural selection, when, from a mass of combinations of organisms arising through symbiosis, only the most adaptable system, the consortium, remains.

Attempting to sketch the historical character of symbiogenesis, Kozo-Polyansky proposes that symbiosis in the narrow sense be differentiated from symbiosis in the broad sense. He understood symbiogenesis in the narrow sense to be a category of hereditable changes that emerge through the natural synthesis of two or more heterogeneous organisms. In this sense, symbiogenesis is a special case of morphogenesis in nature. In its broad sense, symbiogenesis is a protracted historical process in which a complex organism or physiologically integrated consortium is formed through the natural selection of biologically beneficial symbiotic combinations. Kozo-Polyansky insisted on delineating the narrow and broad senses of the concept of symbiogenesis. Mixing them would mean identifying the evolutionary process with the process of inherited variation, which would be paramount to introducing a mechanistic principle to evolutionary study (Kozo-Polyansky 1932, 43).

Kozo-Polyansky thus devoted great attention to theoretical aspects of the concept of symbiogenesis. Approaching the treatment of symbiogenesis from the theory of natural selection, he substantiated a Darwinist understanding of the evolutionary significance of symbiosis. He expressed this understanding by recognizing symbiogenesis as a means of evolutionary transformation that is driven by natural selection. Stating the concept of symbiogenesis from the Darwinist position is Kozo-Polyansky's great contribution.

A New Principle of Biology: An Essay on the Theory of Symbiogenesis (1924)

By attributing great significance to the theoretical analysis of symbiogenesis, Kozo-Polyansky achieved a great deal toward this end. He also devoted great efforts to substantiating the concept with fact. Kozo-Polyansky felt that the immediate task in elaborating symbiogenesis consisted in collecting as much factual data as possible to demonstrate the process of symbiogenesis. Only thus "could [the principle of symbiogenesis] attract the attention of scholars and become the property of wide scientific circles" (Kozo-Polyansky 1924, 136). His work *A New Principle of Biology: An Essay on the Theory of Symbiogenesis* (1924) was devoted to this problem. By dedicating the book to Famintsyn, Kozo-Polyansky emphasized the continuity in the development of symbiogenesis.

Kozo-Polyansky concentrated on several categories of facts, the first relating to symbiotic processes among the most primitive organisms. To this group, which he called *cytodes,* he relegated bacteria and blue-green algae (Cyanophyceae [cyanobacteria]). They have "no nucleus, which is always present both in plant and animal cells, no plastids (in particular no green chloroplasts), so characteristic for plant cells, no cell center, typical for the animal cell, and such 'organs' [organelles], the presence of which is inextricably linked with the concept of the cell" (ibid., 9). He therefore proposed to designate the structural organization of cytodes with the special term *bioblast.** Bacteria and blue-greens are not only the most simply organized forms but also represent the most ancient type of organisms—*primary organisms* of sorts. Finally, they possess a characteristic peculiarity: They are highly disposed to a "social" way of life characterized by an inclination toward forming zoogloea [flocs], as well as colonies and consortia, varying both in composition and in complexity.

The capacity of cytodes to form consortia is of special interest to the theory of symbiogenesis. Zoogloea consortia can serve as good models of real [eukaryotic—L. Kh.] cells. "Studying the consortia of cytodes, we are present at the formation of a model of the cell. . . . The study of cytodes in their symbiotic life leads to the recognition of their participation in the formation of the cell in

---

*The term *bioblast* was suggested by Kozo-Polyansky even before the introduction into scientific usage of the concepts *prokaryote* and *eukaryote* (which were used to designate organisms on different levels of cellular organization) by E. Chatton. According to their organization, *bioblast cytodes* are prokaryotic microorganisms [bacteria].

general, as well as their part in the role of its organelles" (ibid., 25). Kozo-Polyansky offered many examples of zoogloea-consortia in nature. He examined symbiotic systems of cytodes of various compositions—according to the extent to which physiological interrelations between partners are learned and according to how they are transferred from generation to generation. He emphasized that all of these zoogloea consortia primarily were described as independent organisms, and only later, after careful research, was their symbiotic nature revealed.

Kozo-Polyansky described the soil zoogloea studied by S. Vinogradsky as an example of the symbiotic combination of bacteria, a well-researched physiological relation between partners that plays an important role in the fixation of free nitrogen (atmospheric N_2). This consortium includes two types of confervoid/filamentous bacteria and the bacterium *Clostridium pasteurianum*. The capacity to fix nitrogen is limited only to clostridia—to typical anaerobes. Favorable conditions for its development are created by its coexistence with aerobic bacteria. The aerobic bacteria themselves receive nitrogen compounds from their partners. Mutual benefit thus lies at the base of this symbiosis.

The consortium *Chlorochromatium aggregatum* was an especially telling example of symbiotic combination of cytodes in a morphologically and physiologically single system, precisely modeling a single-celled eukaryote. Spindle-shaped bacteria with one long flagellum occupied the central place in this consortium. A large number of rodlike green bacteria, resembling the free-living chlorobacteria *Pelodictyon*, were present in its mucous membrane in longitudinal rows. This consortium was originally described by R. Lauterborn in 1906 as one speckled bacteria (*Chlorobium mirabile*), and only in 1914 did I. Buder succeed in establishing that this was not a single bacterium but rather a whole consortium.

The consortium *Pelochromatium roseum* has a similar history. It was first described by Lauterborn as a single bacterium. Currently, this type of consortium, of significance from an ecological viewpoint, has attracted the attention of researchers (Kuznetsov 1974). In the center of the consortium *Pelochromatium roseum* is located a motile bacterium, related to sulfate-reducing bacteria. On its membrane are photosynthetic green sulfur bacteria. Located in the photic zone of lakes, these bacteria supply the central sulfate-reducing bacteria with organic substances. The central bacterium, manufacturing hydrogen sulfide in the microzone of the same consortium, supplies photosynthesizing bacteria with a hydrogen donor. Such relations between organisms in consortium benefit the system as a whole, since they create favorable conditions for the development of *Pelochromatium roseum* in lakes, "where in the anaerobic zone the hydrogen sulfide is distinguished through analytical methods only in traces" (ibid., 22).

Kozo-Polyansky presented many examples of zoogloea consortia, which were researched by Pascher, Buder, Perfil'ev, and others. Examining these data, he concluded: "Every such total unit in relation to structure and vital functions presents something incomparably more complex and hardy than each of its partners, taken separately. The formation of such zoogloea can be examined as an example of evolution through symbiogenesis—the origination of more 'complete' organisms from simple ones, by means of natural synthesis" (Kozo-Polyansky 1924, 21).

The second group of facts relates to the cell and its organelles. Data substantiating the physiological independence of cell organelles on the one hand, and their similarity to cytodes on the other, caught the attention of scholars.

Kozo-Polyansky emphasized that the conviction of the endogenous origin of green and yellow-green nuclei, located in the cytoplasm of several species of marine animals, persisted for a long time. Only at the beginning of the 1870s was it demonstrated by the work of L. Tsenkovsky that yellow corpuscles of radiolarians were suited to exist outside of the host cell and to reproduce vigorously after leaving it. The "nuclei" [centers] of the radiolarians are whole cells with membrane, nucleus, and plastids of their own. Their genetic continuity through a series of generations was demonstrated. In 1881–83 the works of Entz, Brandt, and Geddes and O. Gammann definitively established that so-called chlorophyllic centers of invertebrates are independent algae leading a symbiotic existence in animal cells. Single-celled algae living in symbiosis were called zoochlorellae and zooxanthellae by Brandt; in 1890 zoochlorellae were identified with *Chlorella vulgaris* and zooxanthellae were identified with algae from the cryptomonad group.

The establishment of the symbiotic nature of these forms clearly indicated that autonomous and even highly specialized organisms, entering into intracellular association, can emerge as cell organelles and that these *organelles*, after a long period, can be seen as products of the differentiation of cytoplasm. "And therefore, the true conception of their nature is initially obscured and gains recognition with great difficulty" (Kozo-Polyansky 1924, 32).

At present, the symbiosis of eukaryotic single-celled algae with heterotrophs is relatively well studied. It has been shown that such association is peculiar to amoebae, ciliates, sponges, Coelenterata, worms, rotifers, mollusks, bryozoans, and others. It is calculated that at least eight different types of algae are encountered as symbionts in more than 150 genera of invertebrates (Raven 1970, 643). The morphological side of mutual relations of symbiont-algae and host-invertebrate has been researched more thoroughly. Thus, for example, it was shown that the green mastigote *Platymonas convoluta*, a symbiont of *Convoluta roscoffensis*, loses its cell membrane, undulipodia, and stigma—structures characteristic of free-living forms (Oschman and Gray 1965).

Kozo-Polyansky discussed the origin of green plastids in plants, noting the absence of gradual transition between chlorophyll, diffusely distributed in blue-greens, and plastids isolated from the remaining plant cell in the cytoplasm. This difference suggested the unique origin of chloroplasts. The absence of gradual transition impeded the argument of chloroplast origin by morphological differentiation. "From the physiological point of view it is not clear what reasons could prompt a similar differentiation, since the presence of chloroplasts before a diffused arrangement of chlorophyll apparently presents no kind of vital advantage," wrote Kozo-Polyansky (1924, 34). He, like Famintsyn, suggested that the decisive argument in favor of the symbiotic nature of chloroplasts could only be their growth in pure cultures. He considered that, thanks to Famintsyn's work, as well as the observations of J. Reinke on chloroplasts in the colorless parts of cells of a rotting pumpkin and the experiments of Lyubimenko on the research of the functioning of isolated chloroplasts in various solutions differing in concentration and composition, plastids "could not fail to be autonomous beings, residing in the cells not unlike chlorellae and xanthellae" (ibid., 38).

Kozo-Polyansky inextricably linked the problem of the origin of plastids to the solution of the question of chloroplast predecessors. In contrast to Merezhkovsky, who saw in blue-greens the predecessors of plastids, he thought that chlorophyll-containing bacteria were original forms into which colorless cells were incorporated and settled symbiotically.

Kozo-Polyansky supported the idea of the symbiotic bacterial origin of the cell nucleus. He considered it possible to discuss seriously the hypothesis of the great cytologist Th. Boveri, according to whom the cell nucleus is a mass of bacteria symbiotically existing in the cell. Fundamental objections to this hypothesis were examined, and attempts were made to find data that could support the colonial and consortial character of the nucleus. The question of the origin of the eukaryotic cell nucleus is one of much current interest, far from definitively resolved. The results of a comparative study of the organization of the genetic material in several groups of bacteria is of special interest (Imshenetsky 1954; Imshenetsky, Zavarzin, and Alferov 1959). The solution to [the question of] how the nuclear apparatus evolved in plants might come from comparative karyology of algae, in which the gradual complexification of the genetic system is clearly traced (Sedova 1977). The karyology of algae, however, remains a poorly known aspect of algal cytology. Because of this, in judging the origin of the cell nucleus, Kozo-Polyansky's scientific judgment, so necessary for objective evaluation, was compromised.

As for mitochondria, Kozo-Polyansky invokes the opinions of contemporary cytologists who recognized the great similarity of mitochondria to bacteria. Based on this, he concludes that in mitochondria as well "one must see symbiotic cytodes or entire systems of such organisms" (Kozo-Polyansky 1924, 54).

Upon examining the origins of centrioles and the blepharoplasts (associated with the kinetic center of the basal apparatus of mastigotes), Kozo-Polyansky emphasized the possibility that they were homologous structures (centrioles/kinetosomes≡Henneguy-Lenhosek theory) and seeing, both in centrioles and in the blepharoplast, an original but good model of the undulipodium-bearing cytode.[†] "The suspicion of the bacterial nature of these kineplasmatic [motility] organelles . . . is fully warranted," wrote Kozo-Polyansky (ibid.). Accordingly, he paid attention to the greatly similar behavior of the blepharoplast in cell division [e.g., *Trypanosoma*] and to the flagellum-bearing partner in the division of bacterial consortia [e.g., *Pelochromatium*]. He recognized that both the blepharoplast and the flagellated consortia bacteria had similar (motility) functions. Ultrastructural observations have generated data on the structure of centrioles and basal bodies [kinetosomes] that apparently support the "equivalence" of these structures (for greater detail see Margulis 1971a). Concerning the "suspicion" of the symbiotic nature of the blepharoplast, first and foremost we must keep in mind the original qualities of this structure and the level of its genetic and biochemical autonomy. Although data on the presence (in the order Kinetoplastida, Kallinikova 1974) of specific DNA (as distinct from nuclear DNA and similar in its properties to mitochondrial DNA) and an endogenous protein-synthesizing apparatus (essentially of mitochondrial character) may be interpreted, in the author's opinion, as evidence that the kinetoplast [of this particular protist group] is "yet another evolutionary variant of the mitochondrial system" (ibid., 1193), the very fact that kinetoplasts are autonomous is extremely interesting for the symbiogenesis hypothesis.

With respect to the structural and functional characteristics of the separate cell organelles, Kozo-Polyansky concluded that "the cell is a collection of heterogeneous, autonomous, live units, leading a symbiotic type of life" (Kozo-Polyansky 1924, 63), and the organelles of the cell are a result of "the joining and incorporation of previously autonomous and independently existing live units" (ibid., 120).

Finally, the third data group Kozo-Polyansky used for the concept of symbiogenesis relates to the existence in nature of complex, "whole-entity" organisms and the absence in plants and animals of "symbiotissue" and "symbioorgans" [the organs and tissues whose development depends on the presence and metabolism of symbionts]. In considering lichens, Kozo-Polyansky wrote that these complex (by nature), multicellular organisms are complexes of forms physiologically equivalent to any plant. He objected to the interpretation of lichen as simply the sum of an alga and a fungus: "By their chemical composition, by their form, by their structure, by their life activity, and by their [propagule]

[†]An expanded discussion of possible symbiotic origins and relations between kinetosome-centriole organelles is in chapter 9.

dissemination, whole-entity organisms demonstrate new characteristics that by no means characterize their separate components" (ibid., 67). As for the character of relations between the lichen components, he wrote: "Without entering into an investigation of the complex problem of the physiological nature of the relation between the partners in a lichen, it is necessary to admit that, whatever the basis of this relationship, the fact that lichens are whole-entity organisms cannot be subject to doubt" (ibid., 68).

Kozo-Polyansky described sponge-algae (*Spongocladia*), the form of whose body represents a unique, qualitatively new form in comparison with each of the partners (sponge of the genus *Reniera fibulata* and filamentous green alga).

He surveyed examples of such tissue and organs in plants and animals that are, by their origin, bound in the mutual existence and life activity of organism-symbionts. These tissues and organs, with varied structure, status, and functions, he called symbiotissue and symbioorgans. Examples include the mucous glands of aquatic ferns and liverworts. The mucous glands of the leaves of *Azolla caroliniana,* which fix atmospheric nitrogen, are distinct symbioorgans formed from filamentous, multicellular growth of the walls of the gland cavity and the filling of this cavity with extracellular colonies of *Anabaena azollae.* In the thallus of *Anthoceros laevis* the mucous glands are filled with colonies of nostocacean, which, penetrating into the mucous crevices, lead to the expansion of the cavity and the growth of wall cells in branching, multicellular growth forming a distinct, homogeneous, parenchymal tissue. Between the cells are found the Nostocaceae. The glands on the aboveground stems and rhizomes of representatives of all species of the genus *Gunnera* are a good example of symbiotissue in plants. These tissues originate by penetration of Nostocacean (*Nostoc punctiforme*) into the mucous canals of the gland and between the cells surrounding the cortical parenchyma. The leaf glands of representatives of the families Myrsinaceae and Rubiaceae (species *Ardisia crispa, Pavetta angustifolia*), as well as the dichotomously branching coralloid organs, which are modified lateral roots. All genera of cycads belong to this category of examples. Recently, the algal symbionts included in Kozo-Polyansky's examples of symbiotissue have been subjected to ultrastructural studies and comparative analyses with free-living forms by N. Lang (1965), D. Neumann et al. (1970), M. Grilli (1972), and M. Sarà (1971) (cited in Gollerbach and Sedova 1974).

Kozo-Polyansky also presented examples of symbiotissue and symbioorgans in insects. Here, he relied on work completed in the period from 1912 to 1922 by P. Buchner, U. Pierantoni, E. Reichenow, P. Portier, and others. The insect symbioorgans considered by Kozo-Polyansky can be combined into three groups. The first contains organs related to the functioning of the digestive apparatus. The structure of these glands and their ability to ferment is defined by the presence and metabolic activity of symbiotic microorganisms. Members of this group include the glandular epithelia of the midgut of all species of ants of the genus *Camponotus,* the racemouse (botryoidal) organ of the midgut of beetles of the group Anobiinae, the "false yolk" of various types of plant lice, the "abdominal organ" of the midgut of the clothes louse, the digestive glands of the tick *Liponyssus saurarum,* feeding on the blood of lizards, and the esophageal glands of the leech (*Placobdella catenigera*). An example of symbioorgans are the incretional glands of bedbugs. In all of the examples he introduced, Kozo-Polyansky considered the morphological and histological structure of the symbioorgan, its disposition within the body of the insect, and its functional role, as well as how the symbionts infect the reproductive organs of their host and are transmitted from generation to generation.

To the second group of symbioorgans belong various additional organs of the sexual apparatus known in many insects, which are created by the presence of symbiotic bacteria of the type *Bacillus mycoides* and yeast, the multicellular body of the surface mollusk *Cyclostoma elegans* distributed in the spinal region between the kidney and the stomach, and the renal organ in Tunicata of the group Molgulidae. To the third group belong the light organs of several insects, Tunicata, and Cephalopoda. Thus, for example, in the larvae of beetles (fireflies) of the family Lampyridae, the light organs consist of two types of cells: some form an opaque zone, fulfilling the role of a reflector, and others form a transparent zone made of illuminating cells. In 1914 Pierantoni showed that the light organs of beetles are formed by the housing of numerous luminous bacteria; they are *mycetomes*. In Tunicata of the genus *Pyrosoma,* each individual of the colony has two well-defined light organs. These organs, as Buchner has shown, are also symbioorgans: illumination is conditioned by the presence and activity of the symbionts populating these organs. In Cephalopoda, not only the morphological structure of the light organs but also their genesis and the specialization in various representatives of this group, have been studied.

Kozo-Polyansky offered mycorrhizae as the clearest and best-studied example of symbioorgans in plants. He asserted that mycorrhizae are essentially a new morphological entity, created by the symbiosis of the root system of plants with the hyphae of a fungus. This type of symbioorgan is the most widely distributed among various groups of plants and plays an important role in the lives of the plants possessed by these organs, fulfilling a trophic function.

Of all mycorrhizae, Kozo-Polyansky emphasized orchids. He saw an unusual physiological picture in orchids in that symbiosis is a necessary condition both for the germination of the seed and for the formation of the roots and tubers and the stages of flowering. Citing M. Rayner's work of 1915–23, Kozo-Polyansky noted that even ordinary heather is essentially a symbiotic organism, formed from a flowering plant and an [ascomycotous] mycorrhizal fungus.

A great deal of material thoroughly investigating mycorrhizae and confirming the validity of Kozo-Polyansky's view has accumulated. With regard to morphology, mycorrhizae can be endotrophic [VA, vesicular-arbuscular, or endomycorrhizae: fungal wall absent; Hartig nets absent; shrublike hyphae vesicles present; almost exclusively within the root cells], ectotrophic [ectomycorrhizae: fungal wall present; fungal hyphae penetrating the interior of the root between the cells of the cortical parenchyma, forming a net; reduction of the root hairs], as well as ectoendotrophic (ectoendomycorrhizae: both). The fungus [or bacterial]-mycorrhizae formers are now well identified. In trees, they are represented by Hymenomycetes, by the genera *Boletus* [many edible mushrooms], *Amanita* [many deadly poisonous mushrooms, bearing cyclopeptide liver toxins], *Paxillus, Russula*; in heathers, by imperfect fungi of the genus *Rhoma*; in orchids, by representatives of the genus *Rhizoctonia* and in several types by basidiomycotes. Mycorrhizae acquire an exceptional significance in the metabolic processes of both the fungal component and the root cells of the host (Rubin and Artsekhovskaya, 1968). Experiments have defined the relations between the components of mycorrhizae in trees and have established the useful role of mycorrhizae-forming fungi in the metabolism of these plants (Shemakhanova 1962). Finally, an attempt to establish the evolutionary path and mechanism of origin of mycorrhizae has been undertaken (Gorbunova 1956). The close mutual conditioning and evolutionary interaction of the mycorrhizal fungus and the plant together constituted a decisive factor in the origin of mycorrhizae. The culmination of this process was the formation of "mycorrhizal associations." One may agree with Kozo-Polyansky's assertion that all the data revealing the characteristics of mycorrhizae support the role of symbiogenesis in the formation of plant symbioorgans.

Kozo-Polyansky concluded that the evidence leads "inevitably and decisively" to symbiosis as a mechanism of evolution. Symbiosis really provides a means for the origin of new traits and structures, the adaptive value of which is defined by natural selection.

In *A New Principle of Biology*, Kozo-Polyansky addressed the question of the significance of symbiogenesis. First, he considered that this concept allows a rational explanation for the origin of cells that contain many membrane-bounded organelles. If one considers the cell as a system of "autonomous, self-sufficient bodies that only to a certain extent have lost their independence" (Kozo-Polyansky 1924, 120), then one may acknowledge that it originated by the synthesis, adhesion, and incorporation of previously independent living beings. "It was not the striving to the division of labor that led to the formation of organelles, but it was the joining into a system by the partners that guaranteed a given program for the division of labor" (ibid.). According to Kozo-Polyansky, symbiogenesis shows clearly how the nature and origin of symbioorgans and symbiotissue in multicellular organisms should be interpreted. It allows one to speak of a multicellular organism as a system or consortium of heterogeneous, elementary, "bioblastic, single- and multicelled organisms" (ibid., 121) and therefore leads to the acknowledgement that the union of heterogeneous entities into a single whole was decisive in the origin of multicellular organisms.

Symbiogenesis acquires further significance for the elaboration of several general biological problems. Kozo-Polyansky thought that it more completely revealed the means of hereditary variation. In addition to the most widespread mechanisms of hereditary variation—mutation and recombination—hereditary variation may be established by *integration* (ibid., 124)—that is, by the natural synthesis of heterogeneous organisms without any sexual process. This category of hereditary variation relates to organisms lacking sexuality that multiply asexually.

Kozo-Polyansky believed that the concept of symbiogenesis makes it possible to pose anew the question of phylogenetic kinship and, in particular, of the reality of "missing links." In those lineages that proceeded by symbiogenesis, the search for intermediate forms is useless, since between the system resulting from the integration of elements and the elements themselves there cannot exist transitions (ibid., 128). Phylogenies are represented not only in divergences of lineage but also in their [anastomosis] converging and merging. However, the union, the integration of several types into one on the basis of symbiosis, as Kozo-Polyansky insisted, should be precisely distinguished from "converging" in the usage of N. Ya. Danilevsky and L. S. Berg, where they have in mind the complete similarity in evolution of several forms belonging to various genetic orders.

In conclusion, *A New Principle of Biology: An Essay on the Theory of Symbiogenesis,* is a significant landmark in the development of the concept of symbiogenesis. Containing a large collection of facts attesting to the validity of the idea of the evolutionary significance of symbiosis, it confirmed well the reality of symbiogenesis itself. Kozo-Polyansky was able to discover facts and observations that revealed the evolutionary means of the symbiotic transformation of organisms.

Conclusion

Kozo-Polyansky (1925a, 13–14) asserted that any concept of evolution needs not merely evidence of evolution by some mechanism but an explanation of the causes of the adaptations. Symbiogenesis resolves this central problem simply. Since evolution was considered to be a process independent of environmental conditions, symbiogenesis acknowledged that this process occurs by the natural selection of useful variants from a mass of random (that is to say, adaptively indirect, or subtle) changes. The concept of symbiogenesis was unique in that, at the same time as it resolved the question of the evolutionary causes in complete accord with the principles of Darwinism, it asserted that the coexistence of organisms (symbiosis) was a factor of evolution. Kozo-Polyansky saw a particular form of the evolutionary process occurring by selection in which the natural synthesis of organisms into a single symbiotic system within a system of interacting factors acquires fundamental significance.

The acknowledgement of natural selection as the primary mechanism of the development of the organic world and the understanding of symbiosis as one of the factors of evolution also formed, according to Kozo-Polyansky, the essence of the symbiogenesis theory as one of a variety of evolutionary concepts within Darwinism (ibid., 22). "The theory of symbiogenesis is a theory of selection relying on the phenomenon of symbiosis" (Kozo-Polyansky 1932, 25).

The connection between symbiogenesis and the Darwinism of natural selection seemed to him so close that, in the event of the demise of the selection doctrine, the theory of symbiogenesis would fall as well (Kozo-Polyansky 1926, 252). As a result, Kozo-Polyansky's interpretation of the significance of symbiosis in the spirit of Darwinism received sharp criticism from Berg (1922).

Kozo-Polyansky objected decisively to symbiosis as a driving force of evolution. He persistently emphasized that his assertion of the universality of symbiogenesis must not be interpreted as a recognition of symbiosis as a universal mechanism of evolution, equal to the influence of natural selection.

Thus, Kozo-Polyansky's main accomplishment in tackling this question was to unite the symbiogenesis hypothesis with Darwin's teachings by considering symbiosis to be a factor of evolution and acknowledging natural selection to be its driving force. Whereas Famintsyn considered the proposition of symbiogenesis to supplement the teaching of Darwin, and Merezhkovsky believed that the theory of symbiogenesis did not agree with the Darwinian solution of the problem of the origin and evolution of organisms, Kozo-Polyansky rethought the concept of symbiogenesis as consistent with the principles of Darwin's teachings.

Finally, one concludes that Kozo-Polyansky deserves great credit in elaboration of the concept of symbiogenesis. His works have been regarded essentially as a new stage in the development of the symbiogenesis hypothesis—the stage of the investigation of the evolutionary role of symbiosis based on Darwinism.

6

A TTITUDES

TOWARD SYMBIOGENESIS

FROM THE 1920s TO 1940s

Perspectives

A great upsurge of interest in symbiogenesis occurred in the 1920s. Judged by the number of publications and the character of the discussion, an abatement began in the 1930s and grew significantly stronger in the 1940s. In the 1950s symbiogenesis was almost never discussed in the Soviet literature.

Many biologists merely mentioned the question of the role of symbiosis in evolution, the majority limiting themselves to general evaluations or expressions either for or against the evolutionary significance of symbiosis (Berg 1922; Ryzhkov 1924, 1927; Keller 1933, 1935; Komarnitsky 1947). Only a few scientists explicitly worked on symbiogenesis. The views of Elenkin and Kozo-Polyansky, who conducted detailed analyses in the 1920s and 1930s on the basis of their own material or the study of data from the literature, have been discussed in chapters 4 and 5. Here the views of the outstanding cytologist-morphologist S. G. Navashin and the evolutionary biologist A. G. Genkel', as well as those of a group of plant physiologists (V. N. Lyubimenko, A. N. Danilov, and P. A. Genkel'), are examined.

Three basic viewpoints among the various ones concerning symbiogenesis can be isolated. Berg (1922) and Keller (1933, 1935) assumed extreme, contradictory positions in evaluating symbiogenesis. A. G. Genkel' (1921, 1923, 1924) expressed (in my opinion) a more cautious and objective point of view. In *The Theories of Evolution*, Berg evaluated the symbiogenesis theory. The Darwinian tendency in symbiogenesis did not suit Berg. He felt that [the theories requiring] a struggle for existence and natural selection as driving the force in evolution were groundless. The symbiogenesis theory, wrote Berg,

> in essence may explain evolution to us just as little as any other theory demanding the acknowledgment of the principle of a struggle for existence. Since selection, as we can now confirm, does not at all select "the most adapted forms," the mutation theory and the theory of hybridization and the theory of symbiogenesis all fade away. Since it is not selection that determines who survives and who perishes, since the life of a species is regulated by certain laws, all the aforementioned structures cannot pretend

to the significance of a theory of *evolution*. Evolution proceeds above them. (Berg 1922, 87)

This criticism acknowledges symbiogenesis as a variety of Darwinism. Rejecting Darwinism, Berg "automatically" rejects those concepts that he determined were based on the selection principle. Berg simply did not consider arguments in favor of symbiogenesis.

Although positive, the evaluation of the symbiogenesis theory found in Keller's works (1933, 1935) is just as extreme. He not only accepted the idea of symbiogenesis, but also categorically universalized it. Plant and animal evolution occurred by the most widespread symbioses, which included distinct phenomena and tendencies of development: community integration, parasitism, lichen symbioses, mycorrhizae, and so forth. "Every plant and animal is a part of a general symbiosis," Keller wrote (1935, 430). Not just the cell, but also its organelles, he considered to be results of symbiogenesis. "The nucleus of the cell, some time in the ancient history of its origin, was once a colony of elementary, living units, the remains of which are bacteriophages and genes—just as multicelled organisms were once colonies of cells" (ibid., 321). Living material, colonies of which formed the origin of the nucleus, was—in its level of development—related to bacteria, in Keller's opinion. Chloroplasts also "were once independent, living units, more simple than the cell" (ibid., 322). Even genes are the remains of ancient, primary creatures, "greatly changed and reworked in a new system" (Keller 1933, 78). Expanding the principle of symbiogenesis to all levels of the organization of life, from the biosphere and its communities to subcellular structures, Keller considered symbiosis a universal factor of adaptive evolution.

Neither Berg nor Keller evaluates symbiogenesis objectively. The former's criticism is indiscriminate negation, lacking analysis of data either for or against the hypothesis. The latter considers one factor of evolution, in this case symbiosis, as a universal mechanism and fails to place it in perspective.

Symbiogenesis was evaluated interestingly by A. G. Genkel', a professor at Perm University (1921, 1923, 1924), who had considered symbiosis by 1901. He studied lichens, *Xanthoria parientina* and *Sticta pulmonacea*, but published the article "On Helotism in Lichens" (1923) much later. Genkel' concluded that the relation between lichen components was helotistic. The term *helotism* was proposed in 1895 by E. Varming to describe the relations between the components in lichen symbioses. According to Varming, in this form of relationship it is as if the fungal component "holds the alga prisoner, consuming its offspring." This type of relation is distinguished from parasitism by the length of time of the process of influence of the fungus on the alga. Supporting the idea of dynamic equilibrium in the lichen symbiosis, Genkel' thought that "both parasitism and endosaprophy and the two types of helotism (passive, when the gonidia develops in the lichen in all stages of its life; and active, when the fungal part actively catches its future component) all probably have a place in nature" (1923, 64).

Genkel' was one of the first to explain from a Darwinist viewpoint the reasons for the origin of a close, joint existence of organisms. He wrote: "Symbiosis is explained by the theory of selection, which is sometimes (incorrectly) called the struggle for existence. This struggle, selecting the most fit, forces them to adapt their organization one with another and to form an association" (Genkel' 1904, 1279).

Genkel' expressed his attitude toward symbiogenesis more definitely much later (in 1923–24)—that it is a stage in the development of lichens. A new, higher phase of association is created in symbiogenesis: the individuality of the components is lost, and a new, complex individual is created (Genkel' 1923, 60). Symbiogenesis was understood as "symbiosis of the highest degree, creating an individual of a new and higher order" (Genkel' 1924, 563).

In evaluating symbiogenesis, Genkel' was correct and responsible. "We will not be severe and fault finding. For us the idea is important, and this idea, as a working hypothesis," he wrote, "can be highly fruitful, especially if it is solidly worked out and explicated" (ibid., 561). Confirming his evaluation of symbiogenesis, Genkel' claimed that it is similar, although indirectly related, to his own views. The symbiotic origin of the green cell was an alluring concept for Genkel' but hopeless from the viewpoint of direct experimental confirmation. He wrote with admiration of the enormous persistence of Famintsyn, who over the course of half a century tried to find evidence in favor of this idea, almost daily creating experiments of isolating and attempting to grow chloroplasts (ibid., 560).

The Position of Navashin

Symbiogenesis also attracted the attention of S. G. Navashin (1916, 1926). In evaluating the data, he did not deny the scientific significance of the concept but thought that it might serve to explain the evolutionary origins of several types of plants—for example, lichens and certain algae (1916, 34). "Since we acknowledge lichens as an independently preserved continuous type," he wrote, "we must not be surprised if several other biologically similar types are discovered among those organisms that are considered "simple" green plants. . . . We have a right to expect that some variety in this regard will be shown by the plants of so diversely composed a group as the algae. It is entirely possible that among them will be found creatures of complex origin, arising, like lichens, from two or perhaps more heterogeneous rudiments" (Navashin 1926, 33).

He also categorically rejected the expansion of the limits of applicability of the symbiogenesis hypothesis and its use as a universal theory of evolution. The discoveries of new, complex organisms would be an extremely important confirmation and extension of the symbiogenesis theory, he wrote, but "they still would not permit us to expand the theory to all green plants as a theoretical solution to the problem of the synthesis of organisms" (ibid., 33).

Navashin rejected the possible explanation of the evolutionary origin of the cell by symbiogenesis. "I have been and remain of the opinion that symbiosis gives us nothing capable of conveying us along the true path toward solving the problem of the origin of living creatures or even of green cells, to say nothing of the problem of the origin of life" (ibid.). A hypothesis of the origin of the cell must introduce an explanation of the origin of two elements: cytoplasm and nucleus. The weakness of the hypothesis of the symbiotic origin of the cell lies in the fact that it proceeds from the facts of the symbiosis of creatures, the nature of which in free existence manifest a definite complexity (Navashin 1916, 34). Navashin himself was inclined to think that the presence of a nucleus and protoplasm (or, as he wrote, "polarities") in the cell is primary and must be taken as data (ibid.).

Navashin's attitude toward symbiogenesis is defined by his understanding of the role of sexual processes in evolution. He recognized the evolutionary significance of sex. Because of sex an entirely new principle of animate organization arose, characterized by the alternation or change of generations. This principle of organization lay at the basis of the evolutionarily progressive initial ancestor of the plant kingdom. A different principle, not connected with sex or the sexual process, lies at the basis of lichen organization and other examples of integrated systems in which close association of two organisms occurs to form one indivisible individual. Navashin emphasized this as the primary distinction between the organization of symbiotic organisms and the "dual organization of other living creatures, which was discovered by studies of the alternation of generations" (ibid., 30). According to him, the absence of the alternation of generations connected with sex made lichens and other forms with similar types of organization dead-end branches and defined their "immobility" in evolution. These arguments precluded his support of symbiogenesis (ibid., 31). Sex and the sexual process, the essence of which consists of various combinations of material elements inherited from both parents, must be recognized as the most powerful of the evolutionary factors, on the broadest scale (ibid., 34–35).

Certainly the origin of the sexual process was one of the important evolutionary innovations, which defined the direction of the progressive development of plants. However, along with this innovation, which has the significance of a universal adaptation, arose others that led to significant biological progress. At the basis of one means of transformation may lie a factor always intensified by the integration of associated organisms.

Navashin's attitude toward symbiogenesis, in particular the symbiotic origin of plastids, is shown in his attitude, formulated in his early work, toward the continuity of plastids (Navashin 1916). The entire question of the origin of plastids in angiosperms was unresolved. Plastid origins could be explained only after an explanation of the problem of continuity—that is, the division and transmission of plastids from one generation to another. The assumption of the continuity of plastids required more precise elaboration of their behavior and fate in fertilization and embryo formation. It was necessary to explain "from where arise those plastids that undoubtedly are found in the tissue of the developing embryo and that later, turning green, turn into chloroplasts" (ibid., 6). Navashin denied the cytological continuity of plastids. The absence of plastid continuity was confirmed by data on the formation of plastids from submicroscopic cytoplasmic particles similar to chondriosomes (mitochondria). He wrote: "In the cytoplasm of the rudiments, by nature indistinguishable from chondriosomes, under the double influence of the plasma and the nucleus arise plastids, which are transformed into the green organs in chloroplasts" (ibid., 13).

The absence of plastid continuity came from an idea about the leading role and participation of the nucleus in all intracellular processes. Inherited plastid traits are formed and transferred to the descendants by the genes of the nucleus, thought Navashin. "We consequently have two theses that now completely change our view of the nature of the chromatophores, in particular, of plants: (1) these structures participate in the process of the ontogenesis of plants in the same way as the various uniform elements of the cells of animals; (2) their properties, as experimental genetics attests, in the process of inheritance are regulated by influences proceeding from the nucleus" (ibid., 15). As E. M. Senchenkova has correctly noted (1973, 49) in defending the thesis about the origin of plastids de novo in the process of cytoplasmic differentiation under nuclear influence, Navashin naturally did not acknowledge any individuality (that is, autonomy) of the plastids. In denying plastid autonomy, Navashin naturally could not admit the possibility of their origin "from green organisms, at one time living freely" (1916, 15).

Although the hypothesis of the symbiotic origin of plastids met with a negative reaction from Navashin, he accepted as an important problem the need to establish a hypothesis about the relation between the nucleus and plastids, or speaking in modern language, about the control by nuclear genes (the genome) of the plastid genetic system (the plastome) over the behavior and metabolism of plastids.

The Works of Lyubimenko

Soviet plant physiologists, especially Lyubimenko (1916, 1917, 1923, 1935), devoted significant attention to symbiogenesis. During the first decade of this century Lyubimenko expressed his opinions in the first and tenth chapters of his doctoral dissertation, "Transformations of the Pigments of Plastids in the Living Tissue of Plants" (1916), and in his paper "The Physiological Independence of Plastids" (1917). His opinion on symbiogenesis, however, was argued in greater detail in the 1920s and 1930s.

Lyubimenko's interest in symbiogenesis was connected with his research on the origin and evolution of photosynthesis;* he thought that a correct under-standing of the evolution of photosynthesis depended on understanding the nature and origin of chloroplasts. "The complexity of the problem of the origin of plastids," he wrote, "creates difficulty in the investigation of the historical development of photosynthesis" (Lyubimenko 1916, 369). The first chapter of his dissertation discusses the origin of the photosynthetic apparatus.

Lyubimenko began from the absence of gradual transition between the mor-phology of the photosynthetic apparatus in phototrophic bacteria and blue-greens on the one hand, and in all green plants on the other. If one admits that the evolution of photosynthesis proceeded from bacteria, assimilating carbon diox-ide chemosynthetically, to photosynthetic algae, then—according to the evolu-tion of the photosynthetic apparatus—one must admit the presence of "gaps" (ibid., 174). In bacteria (*Chlorobium, Pelodictyon, Chlorochromatium*), as in blue-greens, the chlorophyll does not reside in plastids; in green algae the plastids are well developed and preserve their morphological traits so well that taxonomy uses them as species-specific traits. This fact contradicts, Lyubimenko wrote, the supposition of plastids "as a differentiated part of the cytoplasm; if such a differentiation were correct, then it is precisely in the lower algae that we would find different stages of it, for—as the example of Cyanophyceae shows—delimi-tation of the chlorophyllic apparatus from the protoplasm of the cell is not necessary for the success of photosynthesis" (ibid., 179).

The absence of transitional forms of differentiated plastids was an important argument against the assumption of their gradual differentiation from cyto-

*Lyubimenko's opinions on the origin and evolution of photosynthesis were considered thoroughly by E. M. Senchenkova (1969), and his evolutionary outlook was covered by K. V. Manoylenko's monograph (1974).

plasm, in favor of a symbiotic origin. Lyubimenko verified that the evidence of symbiosis in plants, and particularly of symbiosis in colored and colorless forms of bacteria, provided only an indirect basis of the idea of plastid symbiotic origins. Direct evidence required the reconstruction, by cultivation, of the original form from which plastids could develop.

The data on the physiological independence of plastids served Lyubimenko as another point of departure in considering the possibility of their symbiotic origin. The cytological observations of plastid formation in the cells of algae and plants by F. Schmitz and Schimper, A. Sapegin's observations of the fate of plastids in mosses at all stages of development, as well as the data on the movement of plastids in the cell, confirmed the hypothesis of plastids as completely independent units in the cell.

Two paths of plastid evolutionary origin were possible. Plastids may have come from ancestors originally existing as free-living organisms, which subsequently entered into symbiosis with the colorless cells of plants; or they could have arisen by the differentiation of parts of the cytoplasm, which originally functioned in photosynthesis. Both of these suppositions required a factual basis. Science had no reliable facts by which one could witness a gradual transformation from freely existing organisms to a symbiosis in which these organisms adapted themselves to one another such that they significantly lost their individuality. This circumstance prevented Lyubimenko from accepting the symbiogenetic origin of plastids.

Noting the cytological data as evidence for the independence of plastids, Lyubimenko thought it necessary to confirm their physiological isolation from the cytoplasm (Lyubimenko 1917). He proceeded from the concepts of plant physiology. The ability of plastids to accomplish photosynthesis demanded this, since the establishment of the limits of the independence of plastids could provide the key to an understanding, as he wrote, "of the internal conditions" of photosynthesis.

Investigating the interaction of plastids with the cytoplasm involved two methods: studying the influence of various conditions on plastid-nuclear relations in cell development, or isolating the plastids and culturing them in artificial media (Lyubimenko 1917, 49). Because methodological difficulties prevented investigation of the very first stages of plastid development, Lyubimenko attempted artificial cultivation. Leaves of the shoots of wheat, peas, lupine, and tobacco were cut into pieces and placed in test tubes with water. After two weeks the cell nucleus and cytoplasm were destroyed, while the plastids remained undamaged, preserving their internal appearance and color. Then pieces of mucus-containing plastids were transferred into nutrient solutions varying in composition and concentration. After six months the results were compiled. The most definite data resulted from experiments with tobacco leaves: the plastids preserved normal size and color and could not be distinguished from the plastids of living leaves. Neither plastid reproduction nor starch formation, however, was observed—that is, the experiments did not provide irrefutable evidence of the normal functioning of isolated chloroplasts.

Although attempts to culture chloroplasts outside the cell were unsuccessful and direct evidence of the physiological independence of plastids was not forthcoming, Lyubimenko continued to adhere to the hypothesis of their autonomy. In a later work (1935), he considered plastid independence again but from the viewpoint of the genetic continuity of the organelles. Lyubimenko thought that the continuity of chondriosomes (mitochondria) was clear, since they were found in the cytoplasm of germ cells. Plastid continuity in sexual reproduction well confirmed indirect data on the heretability of the various forms of variegated-leaved plants. "Such a transfer," he wrote, "fully corresponds with the hypothesis that plastids are transmitted directly from generation to generation, with all the physiological abilities that characterize them" (Lyubimenko 1935; cited from 1963, 330).

Lyubimenko deserves credit for using physiological data to infer the evolutionary significance of symbiosis (Lyubimenko 1923). He acknowledged that the various forms of physical association play an important role in the growth and development of plants. Considering symbiosis an adaptation of plants, he thought that defining its adaptive significance required investigation of the metabolic process of the associated organisms.

Lyubimenko distinguished between symbiosis in its broad and narrow senses. He included various forms of association, including those in which a physiological relation between the symbionts is absent, for example, when one of the symbionts adheres to the other. He included epiphytes (for example, the algae of the species *Gloeocapsa* and *Trentepohlia;* mosses from the genera *Orthotrichum, Ulota,* and others; and plants from the families Gesneraceae, Orchidaceae, Liliaceae, Araceae, and others), semi-epiphytes (for example, *Carludowica plumieri, Anthurium palmatum,* and the species *Philodendron*), climbing and creeping plants (several species of bedstraw, climbing roses, several species of *Rubus,* the species *Calamus,* hops, bindweed, creeping buckwheat, wisteria, and creeping honeysuckle), as well as various forms of association of plants with animals in the broad sense.

A direct physiological exchange between the symbionts or, as Lyubimenko wrote, "a harmonious combination of physiological substances" (ibid., 923), he included within the narrow sense of symbiosis. The combination of physiological substances into an integrated system leads to the formation of independent biological units or individuals of a complex composition. Lyubimenko considered the loss or severe limitation of the organic capabilities of the symbionts to an independent existence (in natural conditions, they could not develop otherwise than in close unity with physiological interchange) to be the criterion for the origin of such biological unities.

Lyubimenko was the first to try to describe the history of the origin of complex organisms. This was not an easy task, since all forms of symbiosis that would represent the gradual transformation from unstable associations to "the complete fusion into a single biological entity" are now lacking. Parasitism could be the first step on the path of origin of such entities. Parasitism in plants is based on the loss of the ability to synthesize organic material: individuals of one species feed at the expense of living individuals of another species. Plants with different stages of degeneration or the complete loss of the photosynthetic apparatus may serve as an example [necrotrophy]. In these cases the plant-host is not destroyed, Lyubimenko emphasized; it is merely used as a source of organic substances. This form of association occurs by combination of the physiological abilities of the host-plant and its parasite; precisely on this basis arises the gradual transformation to symbiosis in the narrow sense [symbiotrophy].

The next step on the path of unification is one in which the host-plant uses the parasite in its own interests. A direct physiological exchange of substances takes place between the symbionts. The combination of two different species into one individual of a complex composition eventually arises. Similar combinations are represented in lichens, which have numerous forms (ibid., 923–24).

During evolution the combination of fungi with the roots of flowering plants led to the formation of a new independent biological entity: mycorrhizae. In the case of ectotrophic mycorrhizae (for example, in tree species with fungi from the families Agaricineae and Tuberaceae), the relationship between the symbionts is rather weak, the association in its initial stages. In the case of endotrophic mycorrhizae (as in orchids), the relationship between the symbionts is stronger— the symbionts have lost their independence and exist in nature only with one another. These associations are evolutionarily more advanced.

Lyubimenko clearly developed the idea of gradual stages of the evolution of the formation of new biological entities by the symbiotic combination of organisms. He was one of the first to express the profound and correct idea that understanding symbiogenesis required an explanation of these gradual transformations from the combination of organisms within the community to their fusion into a single physiological whole, a complex organism.

Such an approach to the understanding of symbiogenesis involved a Darwinian interpretation. Lyubimenko precisely formulated the question of the necessity of investigating the physiological basis of symbiogenesis. The value of this idea becomes particularly clear in the light of the modern problem of the evolutionary physiology of plants. For example, in the work of K. M. Sytnik and A. V.

Gordetsky (1974), the necessity of comparative physiological investigations to explain the symbiotic origin of photosynthetic and respiratory plastids is emphasized. Attention was directed to the environmental factors conditioning the tendency of algae and fungi to form the lichen symbiosis. The study of the function and reaction of autotrophic and heterotrophic single- and multicellular organisms, related by symbiosis, was added to the number of important problems of the evolutionary physiology of plants.

The Research of Danilov and Genkel'

An original opinion on symbiogenesis was put forth by the botanist-physiologist Afanasy Nikolaevich Danilov (1921, 1927, 1929, 1933), who expressed himself most completely in "Symbiosis as a Factor in Evolution" (1921).

For Danilov, Famintsyn's assertion about the great evolutionary significance of symbiosis was correct. Confirming a role for symbiosis in evolution, Danilov began from the fact that a symbiotic form of the relation between organisms frequently is one of the important conditions in the life of the organism. Physical association, therefore, in all its manifestations, he wrote, cannot be excluded "from the series of diverse conditions in which the evolution of organisms was and is accomplished" (Danilov 1921, 123). The significance of symbiosis was confirmed by the evolution of the mycorrhizae of orchids. The phenomenon of mycorrhizae was one of the most essential conditions of the development of orchids, Danilov asserted, subscribing to the conclusion of N. Bernard (1909). Their evolution involves the working out of adaptations, serving as a more complete symbiosis, and the formation of numerous species is due to the activity of the fungus symbiont. Lichens, so-called witches broom, nodules on the roots of leguminous plants, double-petalledness in flowers, and so on, also confirm the evolutionary significance of symbiosis. These facts attest to the role of symbiosis in plant evolution: "It cannot be doubted that symbiosis is a powerful factor in the formation of new characteristics and morphologies. . . . Symbiosis, undoubtedly, must be placed among the factors of evolution" (ibid., 129).

There can, however, be various interpretations of the evolutionary role of symbiosis. Evolution by symbiosis can be considered a process in which complex symbiotic relations lead to the formation of new symbiotic organisms. This *constructive* (formative) view was held by Famintsyn. Symbiosis can be seen as a factor in the evolution of the traits and abilities of each component, constituting an association. The evolutionary role of symbiosis will be seen by comparing the changes of the traits in each partner with their condition outside the symbiosis. The extent of influence depends on the profundity of the relations between the symbionts (ibid., 123). This outlook held by Danilov, finding persuasive confirmation in lichens, more correctly explains the evolutionary significance of symbiosis. He wrote, "the influence of symbiosis on both symbionts is so significant that each of them forms its own completely new biological race in comparison with how it appears outside of symbiosis. Because of this profound influence of symbiosis on the symbionts, a close bond is created between them, a result of which is the development of lichens—a symbiotic complex" (Danilov 1929, 237).

Emphasizing the exceptional importance of information on lichens, Danilov noted that it was insufficient for solving the question of the evolutionary path of symbiosis. The absence of data identifying the gonidia of each lichen species and ignorance of the "original form" of the fungi that have become components of lichen symbioses allow only guesses on how profoundly each symbiont has evolved under the influence of symbiosis. This limits the evidence for symbiogenesis.

Conducting research toward this end, Danilov claimed that it was necessary to explain the influence of symbionts on one another, their relations, and the influence of their combined existence on the development and morphology of each component to understand the true significance of symbiosis. Using histological analysis of lichen sections from *Evernia prunastri, E. furfuracea, Parmelia sulcata, Ramalina farinacea, Usnea barbata,* and others, he showed that the hyphae of the fungus penetrate the living cells of the algal interior of the gonidia by haustoria, branch between the cell wall and membrane, and permeate the cytoplasm in all directions (Danilov 1910, 47).[†] Danilov concluded that there was a parasitic influence of the hypha on the gonidia and the oppression of the alga by the fungal component. Later, he formulated this position more specifically: the fungus uses the green component as every parasite uses its substrate—as the body of the host (Danilov 1929).

In the lichen symbiosis, the fungus is located in conditions that do not differ significantly from the complex of ecological factors affecting any freely living plant and could be equated with the environmental conditions characterizing parasitism.[§] The influence of symbiosis on the fungal component appears as an intensification of the functional differentiation of the hyphae and in its fertility.

[†]As noted, Elenkin, confirming the scientific value of this data, particularly noted the original character of Danilov's work: "These were the first observations to reveal the haustorium in various stages of development in the gonidia of all types of lichens (fruticose, foliose, and crustose), and they also established the fact of the union of the hyphal net in the gonidia of the observed hyphae" (Elenkin 1910).

[§]In addition to the recognition of lichen symbiosis and parasitism as phenomena of a single order, Danilov wrote about the differences between them. The parasitic character of the interrelations of the symbionts is distinguished from ordinary parasitism by the profound mutual adaptation of the symbionts and is expressed in the development of the thallus. Therefore, the parasitism of lichen symbiosis should be called *mutualistic parasitism* (Danilov 1929, 257).

The influence of symbiosis appears even more significant with regard to the algal component. Investigations of the development and morphology of *Nostoc* gonidia cultured outside of symbiosis showed that in symbiosis many *Nostoc* abilities appeared "suppressed" or did not develop at all (Danilov 1927, 89). In *Nostoc,* living in symbiosis in the thallus of heteromerous lichens, the coccus, which is poorly suited to life outside of symbiosis, is the only form of development that remains. In *Nostoc* living in symbiosis in the *Tricho* crustose lichens, however, the coccoidal stage is rarely seen (ibid., 89). In another case, the alga *Nostoc sphaericum,* from the lichen *Leptogium issatschenkoi,* in its free state forms spherical colonies, while in the lichen is a branched and lumpy form. The explanation of this could be that "the process of germination of the alga does not proceed to completion because of the introduction of the fungus" (Danilov 1933, 40). Danilov thus concludes that the influence of symbiosis on the algal component is great, but it does not differ essentially from the influence found in the external conditions of growth of those algae living outside symbiosis: "Life inside symbiosis is distinguished only by an increase of the sum of external conditions by one, in which all the remaining ones are reflected, and this one of the number of external condition is symbiosis" (Danilov 1927, 30).

As a result of its influence, it does not differ essentially from the activity of the complex of ecological conditions on free-living plants; symbiosis thus can be compared to the influence of light, heat, moisture, and so on (Danilov 1929, 225). As a factor of evolution, symbiosis can be placed, in Danilov's opinion, on a level with light, heat, the moisture of the soil, and so on. Proceeding from this assertion, Danilov concluded that the activity of symbiosis is the same and is subject to the same principle by which external conditions influence the organism—that is, the principle of dynamic equilibrium (ibid., 254). Symbiosis exercises a strong influence on associated organisms, and under its influence morphological or functional adaptations, characteristics of new biological varieties, may arise. The principle of dynamic equilibrium underlies symbiosis as a factor of the evolution of those organisms entering into association.

Thus, Danilov's understanding of symbiogenesis acknowledges the evolutionary transformation of each component in a partnership. His interpretation of the evolutionary role of symbiosis differs from the ideas of Famintsyn, Merezhkovsky, Kozo-Polyansky, and Lyubimenko. The problem as proposed by Danilov merits attention in that his data from lichen research, good material for working out the morphological basis of symbiogenesis, are undoubtedly interesting.

Symbiogenesis was discussed by P. A. Genkel', who was highly critical of the concept, over the course of more than forty years (Genkel' and Yuzhakova 1936; A. G. Genkel' 1938, 1946, 1974, 1976, 1977; P. A. Genkel' et al. 1949). He was interested in methodology. Genkel' thought symbiogenesis was not correct because it is based on incorrect principles (Genkel' and Yuzhakov 1936, 317). One perceives the mechanisms of the concept in the idea that the process of evolution reduces to the simple combination of a series of elemental substances. "Carrying out the theory of symbiogenesis consistently," Genkel' and Yuzhakov wrote, "leads scholars to an extremely idealistic interpretation of the process of evolution as a process of combination and not a process of development" (ibid.). In Genkel''s opinion, the methodological positions from which the concept of symbiogenesis considers the evolutionary process were similar to J. Lotsy's theory.

But the symbiogenesis hypothesis correctly posed the problem of various mechanisms of evolutionary transformation, the complexification of organisms and, further, it approached a truly dialectical understanding of the factors of organic evolution. The factual study of symbiogenesis showed precisely that differentiation and complexification of a system may be attained not only by the isolation of heterogeneous parts within a system, but also by union of formerly independent systems into a single whole without the complete suppression of their qualitative distinctiveness and autonomy (Zavadsky 1967). Genkel''s opinion about the methodological inadequacies of symbiogenesis cannot be acknowledged as correct.

Famintsyn's failure to isolate chloroplasts was a consequence of mistaken methodological principles (Genkel' et al. 1949).[‖] Although he was subjected to Genkel''s sharp (though unsuccessful) criticism, viewed objectively, Famintsyn's works and those of his coworkers made a significant contribution to symbiogenesis. This contribution touches on data that more profoundly revealed the complex nature of lichen symbioses. It was in a lichen that nitrogen-fixing bacteria were first observed and isolated. It was shown that the isolated bacteria fix atmospheric nitrogen in nitrogen-poor environments (Genkel' and Yuzhakov 1936). *Azotobacter* cells were discovered near the green gonidia and on their surfaces in the lichen *Lecanora esculenta* (Iskina 1938). Later, similar data were attained as well in lichens of the species *Evernia furfuracea* Monn., *Lobaria pulmanaria* Hoffm., *Caloplaca* sp., *Lecanora coniza* (Ach.) Nyl., and others (Genkel' 1946). From these studies the nature of symbiosis in lichens was concluded to be more complex than had been supposed: not two, but three components take part—a fungus, an alga, and a nitrogen-fixing bacterium (Genkel' and Yuzhakov 1936).[¶]

[‖]Already in 1937, Kozo-Polyansky had demonstrated that such criticism was unfounded (for more detail, see chapter 5).

[¶]In one of their last works connected with the quantitative analysis of the fixation of nitrogen in lichens (Genkel' and Plotnikova 1973), the questionable nature of the problem of the threefold symbiosis of lichens was emphasized. In natural circumstances, as the authors note, it is extremely difficult to evaluate the true nature of lichens: "the third component does play a notable role, but without question this proposition still cannot be considered finally settled" (ibid., 809).

The results of these investigations, speaking in favor of the threefold nature of lichen symbiosis, were reflected in the concept of *symbiomorphosis*. "We suggest that cases of symbiosis, such as lichen symbiosis, in which an organism is made up of several components and reproduces as a single whole be called symbiomorphosis. This is the most complex type of symbiosis" (A. G. Genkel' 1938, 17). According to Genkel', the physiological basis of symbiomorphosis is the osmotic exchange of the metabolic products of the components that compose it. In a lichen, the fungus—the basic component of the symbiosis—uses the carbohydrates of the alga and ammonia formed by the azotobacteria. The azotobacteria use assimilators, as well as biotic hormones, produced by the green cells of the alga (Genkel' 1946; Genkel' and Plotnikova 1973). A unique exchange of materials between the symbionts leads to a formation that is well delineated from the outside environment and has, as its own internal environment, a closed system: the lichen symbiosis. Precisely the presence of "an internal environment" with the characteristic exchange of materials between the components, as well as the ability of the lichen to reproduce as an integrated whole, allows one "to think that this is symbiosis of the most developed and complex form"—symbiomorphosis (Genkel' 1946, 52).

The idea of *symbiomorphosis* attracted great interest. First of all, it is valuable because it objectively reflects the existence in nature of compound, complex organisms that by their origin manifest the symbiotic form of the integration of several heterogeneous individuals. The basis of symbiotic organisms becomes the presence of a common material exchange between the components and the ability of the complex to reproduce as an integrated whole. The concept of *symbiomorphosis* is also valuable because it leads to the recognition of a path of evolution in which symbiosis is a leading mechanism. An understanding of symbiomorphosis as a phenomenon of the evolutionary formation of a living system in which symbionts form their own internal environment with a characteristic exchange of materials and an ability to multiply as an integrated whole, helps to acknowledge the reality of symbiogenesis, even though the author of this concept did not himself belong to the supporters of the symbiogenesis hypothesis.

Genkel' evaluated symbiogenesis again in later years. In his introduction to A. Frey-Wyssling's "The Comparative Organollography of Cytoplasm" (1976), he wrote that the works of Frey-Wyssling (completed in cooperation with K. Mühlethaler), and research of Mühlethaler and P. Bell were against the proposition of the symbiogenetic origin of chloroplasts and mitochondria. "It should be noted that the symbiotic hypothesis," Genkel' wrote, citing Kursanov and Komarnitsky's book (1945), "is built on analogies and can hardly be considered proved to any extent" (Genkel' 1976, 6). In reality, based on studies of oogenesis in the fern

Pteridium aquilinum, Mühlethaler and Bell concluded that all plastids and mitochondria are destroyed upon the maturation of the ovule and later form de novo from the nucleus as a result of the evagination of the nuclear membrane (Mühlethaler and Bell 1962). The buds formed in this way, separately from the nucleus, begin to develop organelles—plastids and mitochondria.

A thorough analysis of the formation of plastids de novo from the nucleus—including the hypothesis of Mühlethaler-Bell—was made by G. Ya. Zhukova (1975) on the basis of wide embryological material. The proposition of Mühlethaler and Bell contradicts the long-known presence of plastids in the cells of the female gametophyte, in particular in the ovules of the cells of *Hyacinthus non-scriptus (Endymion non-scriptus), Daphne blagayana, Torenia asiatica, Atrichum undulatum, Anthoceros laevis* (Lyubimenko 1916), and a series of other plants, as demonstrated by K. Noack in 1921, O. Renner in 1934 and 1936, L. Anderson in 1936, K. Steffen in 1951, and F. Schötz and V. Stubbe in 1962 (Zhukova 1975). Moreover, Bell and Mühlethaler's data are doubted because of their incomplete methods. Moreover, observations of plastids at the formation of gametophytes in mosses, as well as electron-microscopic study of the organelles of the ovules of a series of flowering plants (for more detail, see Zhukova 1975), do not confirm these authors' conclusions. Zhukova wrote: "There is a series of serious objections against the hypothesis of Bell and Mühlethaler. The alterations of plastids observed by them affect their internal structure; it becomes simpler and the lamellae disappear. These changes, which Bell and Mühlethaler consider degenerative, may well be interpreted as processes of the reorganization of plastids into simpler forms, taking part in the transfer of genetic material to the new generation. Another mechanism, for example, one involving the lysosomic structure, would be more effective for the elimination of these organelles. The described process appears to be a process of dedifferentiation" (ibid., 720).

7

CONTEMPORARY CONCEPTS OF SYMBIOGENESIS

The modern study of symbiogenesis, the evolutionary significance of symbiosis, began in the 1960s; data from various fields of biology were
examined in the light of knowledge of mechanisms and laws of evolution. Data
considered to be proof of symbiogenesis were gathered from lichenology, algol-
ogy, protistology, cytology, molecular biology, virology, genetics of bacteria, and
other fields of contemporary biology. Classical and new physical methods (such
as electron microscopy, analytical centrifugation), as well as the improvement of
biochemical, cytophysiological, and other research methods, were used.

Proof of Symbiogenesis

The origin of a large higher taxon (of plants), lichens, serves as irrefutable proof
of symbiogenesis. Extensive research on these organisms has shown symbiogenesis in lichens to be a fact and not merely theoretical assumption. Data detailing
the evolution of gelatinous lichens are of particular value (Elenkin 1912, 1922a,
1922b, 1922c; Danilov 1929; Gollerbach 1928, 1930; Gollerbach and Sedova
1974), as discussed in chapter 1.

The emergence of mycorrhizae, a complex life form created by the symbiotic
unification of fungal hyphae with roots of plants, is a well-studied fact attesting to symbiogenesis. Mycorrhiza data, including the facts upon which Kozo-
Polyansky based his concept of symbiogenesis, are in chapter 5.

The existence in plants and animals of many symbioorgans and symbiotissues
(that is, the organs and tissues whose emergence hinges on the presence and life
activity of symbionts) supports the reality of symbiogenesis—for example, the
rumen and hypertrophied hindguts of animals that feed on cellulose, seeds, and
so forth, blood-sucking insects, and the bacterial organs of luminous animals.

The presence of symbioorgans and symbiotissues in plants was discussed in
chapter 5. An exhaustive summary of symbioorgans in animal taxa is provided
by Buchner (1965) From morphological, histological, and functional studies of
endosymbioses in animals, Buchner described symbiont transfer from generation
to generation. Buchner's research on the relation between insects and their
microbial endosymbionts showed many cases in which microbes are beneficial to

the insect, Sukhov (1942) noted. Special devices exist that ensure the infection of the larval insect; in other cases the delivery of microbes with the egg cell is guaranteed.

Research on several of the most advanced forms of endocyanosis (endosymbioses of cyanobacteria with colorless protists) belongs to the less well-documented data supporting symbiogenesis. Interest in these comes primarily from the extent of component integration, which attains such a high level that the symbiotic complex transforms to appear to be a single, one-celled organism. As is well known, Korshikov (1924) described endocyanosis of *Cyanophora paradoxa* in this way. [Editors' note: These organisms have been grouped together as Glaucocystophyta of the kingdom Protoctista (see Kies and Kremer 1990, Editors' Introduction).]

Intracellular endocyanosis research is of great interest to problems of plastid evolution. Analysis of the complex features of cyanelles of various endosymbioses and the comparative study of the structure and functional ties of cyanelle endosymbioses could provide a model of the origin of chloroplasts. Indicating the necessity for careful morphological, physiological, and cytological study, Elenkin posed this question in 1936. The relation between cyanelles and host cells has been shown; their transformations into the series of endocyanoses has likewise been revealed: *Geosiphon pyriforme, Glaucocystis nostochinearum* and *Cyanophora paradoxa* (Elenkin 1936; Hall and Claus 1963, 1967; Taylor 1970; Gollerbach and Sedova, 1974).

In the process of the gradual strengthening of relations between components, a systematic enlargement of the whole symbiotic complex and a strengthening of its individuality occurs in proportion to the extent of their union. Comparison of these endocyanoses showed that the development of symbiotic union is accompanied by morphological and functional specialization and subjection of the individuality of symbionts, expressed as the alteration of a series of features of the symbionts compared to the organization of corresponding free-living algae.

The most primitive endocyanosis, described initially as an independent organism, is *Geosiphon pyriforme*. This is composed of colorless, siphonaceous algae, according to von Wettstein (1915), with endosymbionts in its cytoplasm—members of the species *Amorphonostoc,* according to Elenkin (1936). Further research (Taylor 1970) showed that the extent of integration between its components is low; they easily exist outside of symbiosis and can remain separate for a relatively long time. Analysis did not reveal structural change in cyanelles; it affirmed features corresponding to those of free-living cyanobacteria. The continuity of this symbiosis is maintained by resynthesis of host-cyanelle relations in each generation.

The endocyanosis *Glaucocystis nostochinearum,* known since the 1880s as a

Contemporary Concepts of Symbiogenesis

single-celled organism, is more advanced with respect to the extent of integration of its components. Study of *Glaucocystis nostochinearum* revealed both its complex nature (symbiosis of the colorless chlorophyte algae from the family Oocystaceae and rodlike cyanobacteria) and important structural changes in host cell and cyanelle organization. The cyanobacterium underwent such strong transformation that it is still not classified. "*Glaucocystis,*" note Gollerbach and Sedova (1974), "is an example of the establishment under symbiosis of the most close relations between partners, which leave their marks even on their morphology" (1974, 1364) [see Kies and Kremer 1990, Editors' Introduction].

First and foremost, the changes affected the cellular covers of the cyanelle. It turned out that the cyanelle was surrounded by two distinct membranes, the outer membrane being a structural element of the cell-host, and the inner belonging to the cyanelle itself. Such a change in the structural complexity of the cyanelle can be explained only as the result "of the force of natural selection, called forth by symbiosis. It shows a general increase in the dependence of the cyanelle on the host" (Taylor 1970, 50). Strong mutual dependence between the symbionts of *Glaucocystis* manifests itself, as Gollerbach and Sedova (1974) note, as well as functional specialization: the cyanelles use nourishing substances produced by the host cell, and the cyanelles photosynthesize, providing the host cell with products of photosynthesis.

Everything known about *Glaucocystis* speaks convincingly of extensive integration between host and cyanelle; the cyanelle developed in the direction of greater complexity as the *pseudochloroplast* (Taylor 1970).

The next step along the path of increased integration between symbionts and dramatic suppression of their individuality to the developing, unified whole is the endocyanosis *Cyanophora paradoxa,* formed by a pyrrophyta-type cryptomonad and strongly altered blue-green of a species of *Chroococcus*. Neither component was viable outside of the symbiotic complex or grew on artificial media. The cyanelle cell wall is greatly reduced, from four layers to two. The cyanelles have a special central body representing the nucleoid (Hall and Claus 1963). These data prove that, in the process of evolution, stable, highly individualized systems emerge through the strengthening of relations and integration of symbionts.

Research on symbiotic cytoplasmic particles of several strains of *Paramecium aurelia* directly support symbiogenesis. The activity of these particles determines the appearance among their carriers of a series of hereditarily determined adaptations (Beale, Jurand, and Preer 1969). One line of paramecia secretes a special substance into the medium that kills other sensitive strains. T. M. Sonneborn, the first to describe this phenomenon, showed that the capacity to kill (the "killer" property) is determined by cytoplasmic killer factors, which he called "kappa particles." Genetic analysis showed that the activity of kappa appears among paramecia when the dominant nuclear gene K is present. Initially identified by Sonneborn as cytoplasmic particles or plasmogenes, kappa particles turned out to be microbial symbionts. Research by Preer and her colleagues showed that kappa particles have their own DNA, possess a double membrane, and resemble rickettsia (Ball 1969). Several types of cytoplasmic endosymbionts, determining properties of their carriers, are different strains of *Paramecium aurelia*. Molecular and electron microscope data allowed Preer to assign a species name to the endosymbionts (Preer, Preer, and Jurand 1974; Bergey's Manual). Thus, alpha particles are in the species *Caedobacter caryophila*, kappa in *Caedobacter taeniospiralis*, mu particles in *Caedobacter conjugatus*, and lambda in *Lytium flagellatum*.

Important research directly related to proving symbiogenesis concerns the ability of species of toxic invertebrates to retain foreign chloroplasts in the cytoplasm of cells of their digestive system. The captured chloroplasts remain intact and in several instances even retain functional activity. Photosynthetic chloroplasts were found in the digestive cells of the rotifer *Ascomorpha ecaudis*, in the cells of the digestive tract of the mollusk *Elysia atroviridis*, and in four species of mollusks belonging to the order Saccoglossa (Taylor 1970; Pinevich et al. 1974). The chloroplasts of *Codium tomentosum*, captive in the cells of *Elysia viridis*, continued to function and are identical in ultrastructure and color with chloroplasts from free-living *Codium tomentosum* (Taylor 1970). Chloroplasts in the digestive gland of the mollusk *Tridachia crispata* (Trench, Green, and Bystrom 1969) absorb radioactive bicarbonate $HC^{14}O_3^-$ in daylight, as detected by autoradiographic methods.

Chloroplasts and Mitochondria

For arguments in favor of symbiogenesis, data from the most important organelles of the cell, chloroplasts and mitochondria, are of exceptional interest. Recent research has revealed these organelles to be highly integrated and independently organized cell structures, posing again the question of their origins. New data on the properties of chloroplasts and mitochondria permit a more intensive comparison of chloroplasts and single-celled blue-greens with mitochondria and bacteria, revealing an entire complex of similar traits. The discussion of the question of the possible direct evolutionary link between chloroplasts and mitochondria and prokaryotic organisms enables a firmer approach to resolution of the role of symbiosis in the origin of cell organelles.

Chloroplasts and mitochondria represent intracellular systems possessing great autonomy. Cytological continuity through a series of cell generations and the presence of their own genetic apparatus and protein synthetic system are the most important features, demonstrating the individuality and autonomy of both chloroplasts and mitochondria.

Although the formation of plastids by the direct division of predecessors in "lower" plants is considered established, the reproduction of chloroplasts and their transfer from one cell generation to another in angiosperms have not been definitively solved (chapter 3). The cytological continuity of plastids in "higher" plants is not so apparent; further refinement is required. In contrast to algal plastids, plant chloroplasts do not form directly by the division of mature differentiated plastids, but rather they develop from colorless proplastids (Breslavets 1963; Kirk 1970; Senchenkova 1973; Zhukova 1975). Proplastids thus ensure the continuity of chloroplasts in plant meristematic cells. This property of plastid morphogenesis in meristem cells occasioned the refutation of chloroplast succession; the assertion about plastid emergence in the process of ontogenesis anew from cytoplasmic particles cannot be ignored (Aleksandrov 1950; Breslavets 1959, 1963). Electron microscope research on the behavior of chloroplasts during cytokinesis led to attempts to dispel differences in interpretations of their continuity. The proplastid reproduction of chloroplasts of plant meristem cells and the function of proplastids should be viewed as a more complete method of supporting plastid continuity by comparison with nonangiosperms (Zhukova 1975). The fate and reproduction of plastids in sexual reproduction, including the formation of reproductive organs and tissues of plants, is even more complex.

Plastid behavior and ontogeny provide a basis for the cytological continuity of these plant cell organelles, although much is insufficiently researched. At present "the question about the universality of the concept of plastid continuity still awaits definitive confirmation, since the division of plastids and their continuity is difficult to trace in tissue that contains leucoplasts and undifferentiated small plastids (proplastids) in meristematic tissues, as well as in tissues that initiate micro- and macrospores and gametes" (ibid., 717).

The principal fact testifying to the autonomy of chloroplasts is the presence of their own genetic system and protein synthetic apparatus, which distinguish them from the corresponding systems of nucleus and of cell cytoplasm.

The myriad of works using diverse methods (cytochemical, autoradiographic, biochemical, electron microscope) have proved the existence of plastid DNA, which is synthesized and replicated in chloroplasts. Chemical and physical characteristics have been studied, including nucleotide sequences, the quantitative and spatial organization of plastid DNA, and the difference between the properties of plastid DNA and the properties of nuclear DNA (summarized by: Gibor and Granick 1964; Antonov 1969; Beridze and Odintsova 1969; Kirk 1970; Raven 1970; Taylor 1970; Pakhomova 1972, 1974; Svetaylo 1973; Senchenkova 1973; Zhukova 1975; Nasyrov 1975; Odintsova 1976).

The quantity of DNA averages 10^{-15}g to 10^{-16}g for each plastid and constitutes 1%–10 per cent of the total DNA of the cell. Chloroplast DNA, $1-2 \times 10^8$ daltons, has a high molecular weight. The nucleotide composition, expressed in molar ratios of complementary pairs of nucleotides of chloroplast DNA is AT rich. Chloroplast DNA is not bound to histone proteins or attached to membrane. Plastid DNA is characterized by a higher-than-nuclear buoyant density in CsCl gradients. The rapid renaturation kinetics testify to the high level of its inner molecular homogeneity. Chloroplast DNA, compact fibrils of 25–30 Å in diameter, lacks chromosomal organization.

The presence of a chloroplast genetic system in such a highly integrated system as the cell is an amazing fact in itself. It requires satisfactory explanation. The suggestion of the emergence of plastids from previously independent single-celled organisms that later entered into symbiosis is the simplest and most convincing explanation, confirmed by new facts. One can think that "the acquisition of an already prepared photosynthetic system by the colorless proto-eukaryotic cells by entering into symbiotic relations with green prokaryotic forms is a great advantage in comparison with the gradual formation of a new photosynthetic apparatus" (Zhukova 1975, 731).

The ability of chloroplasts to synthesize protein attests to their great autonomy. All have the basic components of a protein-synthesizing system. Research was carried out in the 1950s and 1960s by N. M. Sisakyan and his school (see also: Filippovich, Svetaylo, and Aliev 1970; Svetaylo 1973), which showed that chloroplasts contain ribosomes, polyribosomes, messenger RNA, various types of transfer RNA, and enzymes for the synthesis of protein. Physicochemical characteristics show that they are distinct from the corresponding components of the cytoplasmic apparatus for the cytoplasmic protein synthesis. Chloroplast ribosomes of plants, according to their sedimentation properties, belong to the 70S type, while the cytoplasmic ribosomes are of the 80S type. The sedimentation coefficient for plastid ribosomes of pea shoots equals 69.8S, but for cytoplasmic ribosomes of the same peas it is 79.8S (as a comparison, the sedimentation coefficient of *Escherichia coli* ribosomes and of cyanobacteria equal, respectively, 69.1S and 70.0S) (Svetaylo 1973).

The ribosomes of plastids and cytoplasm differ in their synthesis of protein in the presence of specific inhibitors. Chloramphenicol inhibits the synthesis of protein by plastid ribosomes, but the cytoplasmic system is insensitive. Differences were also revealed with regard to ribosomal RNA. The sedimentation constant of the RNA of plastid ribosomes is close to that of the RNA of bacterial ribosomes and lower than that of the cytoplasmic ribosomes.

Data convincingly attest to the autonomous protein-synthesizing apparatus in plastids, demonstrating its distinctiveness from that of the cell cytoplasm.

In addition to the discovery of plastid DNA, the genetic autonomy of chloroplasts received indirect support in work investigating plastid colors. By 1908–09, Correns had described in *Mirabilis jalapa* the inheritance of color, which did not correspond to Mendel's laws. The non-Mendelian inheritance of defects in plastid color through maternal inheritance attests to the presence of unique genetic factors localized in the chloroplasts themselves. "In this case we have a clear example of inheritance through the cytoplasm of self-reproducing elements (plastids), which influence the plant phenotype and transform one into another (green-white) through mutation" (Dubinin 1976, 423). Cases of the unilateral, nonchromosomal transfer of many traits related to the action of the plastid genes are now known in many species of plants and fungi. The modern understanding of plastid inheritance is described in Jinks 1966; Kirk 1970; Sager 1975; and Dubinin 1976.

The relative genetic autonomy of chloroplasts is a sufficiently established fact, but this autonomy is not complete. The genetic activity of plastids is limited by the quantity of coded information accommodated in plastid DNA. Chloroplast DNA is large enough to provide information for the synthesis of only 100–200 proteins with a molecular weight of 50 kd (Nasyrov 1975a). Genetic research indicates development of the plastid phenotype, and functional capabilities are regulated not only by the plastid factors but are also under the direct control of nuclear genes. Hundreds of chromosomal genes are known whose mutations alter plastid attributes (Dubinin 1976). Nuclear genes may act on the plastids by reducing the metabolite supply or by forming inhibiting or activating substances that act on the plastid genetic system (Gibor and Granick 1964).

The limitations of the genetic autonomy of plastids were excellently demonstrated by Kirk (1970), who showed that genetic regulation of the three most important stages of chlorophyll synthesis is facilitated by nuclear genes. Moreover, data show that some pigments, enzymes, and ribosomal and structural proteins are coded by nuclear DNA (Pinevich et al. 1974).

The interaction of nucleus and plastid genetic systems in the development and functional activity of the photosynthetic apparatus is of great significance in evaluating the extent of plastid autonomy (Nasyrov 1975a, 1975b). *Acetabularia mediterranea* has become a valuable subject of such research. The genetic regulation of chloroplast biogenesis shows clearly in enucleated cells of this alga. The comparative analysis of the rates of division, photosynthesis, and DNA and RNA synthesis in the chloroplasts of enucleated and intact cells, and experiments on induced mutation in chloroplasts, in combination with transcription and translation inhibitor studies, show that the development and functional activity of the photosynthetic apparatus is controlled by the polygenic, balanced system of nucleus and chloroplasts (Nasyrov 1975a, 1975b). The formation of the protein-synthesizing system, which forms thylakoid proteins, is controlled by nuclear genes, while chloroplast DNA determines the genetic continuity of the plastids, codes for all three types of RNA and for some important proteins.

The interaction of two relatively independent genetic systems, nucleus and chloroplasts, within the confines of a single cell and the methods of the harmonious cooperation of the nuclear and plastid genes still remain unresolved. The material presented here supports the hypothesis of plastid symbiogenesis.

Other evidence accumulating in support of the symbiogenetic origin of plastids compares traits of chloroplasts with those of representative cyanobacteria. Similar traits of structural organization and genetic and biochemical attributes exist that have provided bases for the confirmation of their common origin.

The existence of similar features between chloroplasts and cyanobacteria is experimentally confirmed in the works of Beridze and Odintsova (1969), Kirk (1970), Taylor (1970), Pakhomova (1972, 1974), Ostroumov (1973), and Svetailo (1973). I mention only those that particularly clearly illustrate this similarity.

Chloroplasts and cyanobacteria have a double membrane and numerous thylakoids lying adjacent to each other in the stroma. Plastid DNA, like cyanobacterial DNA, is not organized into chromosomes. The DNA of these forms is arranged in small, compact fibrils with a diameter of 25–30 Å (Ris and Plaut 1962; and Werz and Kellner 1968).

The DNA structure is similar in both chloroplasts and cyanobacteria. According to a UV analysis of the nucleotide composition of cyanobacteria DNA from the families Nostocaceae and Scytonemataceae—without regard to the extreme variability of this trait—their DNA, like algal chloroplast DNA, is of the AT type (Pakhomova 1972, 1974). The DNA of chloroplasts and cyanobacteria is characterized by the unimodal distribution in the CsCl density gradient centrifugation. The satellite fractions in chloroplast DNA and in cyanobacteria may be polysaccharide components.

These similarities between plastid and cyanobacterial DNA, shown by modern investigations, support the hypothesis that "a genetic similarity exists between prokaryotic cells and the organelles of eukaryotes" (Pakhomova 1972, 190).

The similarity of chloroplast and cyanobacterial DNA is also seen in physicochemical attributes. The melting temperature curves are quite similar, attesting to extensive intramolecular homogeneity of both of these DNA's. The degree of dispersion (delta 2/3) of plastid DNA preparations from *Pisum sativum* L., *Euglena gracilis* Klebs., and the DNA of *Spirulina platensis* (Com.) Geitl. and *Nostoc* 467 was, respectively, 6.5, 6.2, 6.8, and 6.4 (Lipskaya, Ivanova, and Grabovskya 1974).

The photosynthetic pigments, the carotenoids, are similar. Cyanobacteria, like chloroplasts, synthesize beta-carotene and its hydroxyl derivatives, as well as its complexes with other substances. They lack *alpha*-carotene, however, and its derivatives (Pinevich and Vasil'eva 1972).

A similarity in ribosomal RNA was shown in plastids (of the red algae *Porphyridium* and *Euglena gracilis*) and in cyanobacteria, as well as in bacteria. This structure was not homologous to cytoplasmic ribosomal RNA (Bonen and Doolittle 1975; Zablen et al. 1975).

Chloroplast ribosomes, like those of cyanobacteria, are of 70S, with regard to their sedimentation properties (Pakhomova 1974).

These data on chloroplasts and their similarity to cyanobacteria attest to the idea that chloroplasts are individualized entities within well-integrated systems—plant cells—and also show a profound essential similarity between chloroplasts and cyanobacteria. These facts, which demand explanation, may be understood clearly if one accepts that an evolutionary relation between chloroplasts and cyanobacteria exists, and further that in the course of evolution, a symbiotic plant cell, functioning and developing as an integrated whole, may arise.

Plant embryology data are significant for accepting the symbiotic origin of chloroplasts. Investigations of the ultrastructure of chloroplasts of the developing embryo of angiosperms lead to the conclusion that the embryo's plastids combining traits of chloroplasts and leucoplasts are transitional forms in the evolutionary transformation of the initial, polyfunctional chloroplast into a leucoplast. All known types of pigment-containing and colorless plastids began from chloroplasts. "If this is so, then the very process of the phylogenetic development of plastids might serve as confirmation of their symbiotic origin; moreover, from the position of the genesis of plastids from the confining membrane, it is difficult to explain why the process of the formation of chloroplasts proceeded first, instead of the formation of chloroplasts and leucoplasts simultaneously" (Zhukova 1975, 731).

Various data regarding significant autonomy and similarity to bacteria have been attained with regard to the other important organelle, that is, mitochondria. An evolutionary approach to this data is presented in the works of Leninger (1966); Nass (1969); Beridze and Odintsova (1969); Roodyn and Wilkie (1970); Margulis (1970); Nayfakh (1972); Ostroumov (1973); Zotin et al. (1975); and many others.

Mitochondria possess a unique genetic system, which (by many indicators) is distinct from the nuclear genetic system. Mitochondrial DNA lacks chromosomal organization and has considerable homogeneity, a higher buoyant density, and a synthesis that differs from that of nuclear DNA. Mitochondrial DNA, as distinct from nuclear DNA, has a circular configuration, as is characteristic of bacterial

DNA. The base pair composition is also distinct from nuclear DNA and similar to that of bacteria. Mitochondrial DNA is attached to the internal membranes, not complexed with histone proteins.

Mitochondria also have a unique protein synthesis system. They contain RNA polymerase, messenger RNA, transfer RNA, and small ribosomes (50–60S). Data on the ability of mitochondria to incorporate amino acids into proteins and on the inhibition of mitochondrial protein synthesis independently of cytoplasmic ribosomes also attest to mitochondrial autonomy. The genetic information stored in mitochondrial DNA is sufficient to code for only 5,000 amino acids (Roodyn and Wilkie 1970).

The autonomous genetic regulation of the mitochondrial biogenesis is clearly quite limited. Mitochondria can synthesize only structural proteins of the internal membrane. The dependence of mitochondrial protein synthesis on nucleo-cytoplasmic control has been shown by experiments involving protein synthesis inhibitors, which suppress the ongoing protein synthesis in mitochondrial ribosomes (Pinus 1973).

Data comparing the characteristics of mitochondria and bacteria (Raven 1970; Nass 1969; Ostroumov 1973; Zotin et al. 1975; and others) are extremely important for an elaboration of their symbiotic origins. The well-defined similarities between mitochondria and bacteria are revealed by general morphology, the localization of enzymes, chemical composition, drug sensitivity, nucleic acid structure, and attributes of the protein-synthesizing apparatus (Roodyn and Wilkie 1970). The presence of such profound and numerous similarities establishes a direct evolutionary relation between mitochondria and bacteria, in S. Nass's opinion, and, as Roodyn and Wilkie emphasize (1970, 71), makes probable the hypothesis that "mitochondria arose from some primitive organisms, similar to bacteria, which penetrated into a nucleus-containing cell and became endosymbionts."

G. G. Gauze (1977) has hypothesized an evolutionary origin of mitochondria. In discussing a hypothesis on mitochondrial origins from plasmids proposed by Raff and Mahler (1972), Gauze suggests that mitochondrial DNA is not a descendant of plasmids but rather a reduced descendant of the chromosomal DNA of the preeukaryotic cell. Gauze hypothesizes that, at some stage of cell evolution, repeated nucleotide sequences arose in DNA. Because of structural isolation resulting from the membrane system, copies of the genome differentiated. These copies could, in time, form the beginning of mitochondrial DNA (Gauze 1977, 270); this hypothesis explains why genes for rRNA and tRNA synthesis are present in mitochondrial DNA.

The data on both the individuality of chloroplasts and mitochondria, as well as on their similarity to bacteria, substantiate the recognition of the symbiogenetic origin of these cell organelles; as Dubinin (1976, 433) justly remarks, such material "shows that in their time, the ancestors of mitochondria took root in the cells of eukaryotes as symbionts. Then they began to provide the cell of the host with energy from the molecule ATP and entered into the composition of the cellular plastids. The same is true of plastids. They also possess DNA and ribosomes of a bacterial type. Their origin was possibly bound to the introduction into the cells of eukaryotes of photosynthesizing bacteria."

The correlation of structural characteristics of DNA with the taxonomic status of organisms is also significant for symbiogenesis. A great contribution to the elaboration of this discipline, called genosystematics, was made by A. N. Belozersky and his school. Belozersky wrote of his goal: "The study, in the first instance, of nucleic acids in a systematic, comparative biochemical, evolutionary aspect is particularly interesting and promising from the point of view of recognizing the material bases of the evolutionary process, since it is precisely these unions that are directly linked to inheritance and variability" (1969, 5). Comparative taxonomic investigations of the genome have shown a close linkage between the structure of the DNA molecule and the taxon of the organisms. Taxonomical specification of DNA composition settled the question of the phylogenetic kinship of separate taxa and of phylogenetic relations between representatives of various taxa, and so forth (Antonov 1969, 1974; Belozersky 1969; Pakhomova 1972, 1974; Mednikov 1975; and others). Research on the frequency of complementary pairs of DNA nucleotides in plants, cyanobacteria, and green algae shows that DNA of the first two groups and DNA of green algae belong to different types. The quantitative content of the GC nucleotide pairs in plants equals 42 mol%, that is, DNA of the AT-type. The same type of DNA has been traced to cyanobacteria. In green algae, the base pair ratio equals 61.3 mol%, indicating GC-rich DNA (Belozersky, Antonov, and Mednikov 1972). These data, attesting to the limited homology of these DNA's, allow us to suggest confidently that green algae and plants stem from different evolutionary roots. If this is not so, then this improbability must be conceded: in the process of evolution a sharp shift equal to more than 19 mol% occurred in the nucleotide composition of DNA. It is more realistic, in Belozersky's opinion, to suggest that plants stem from primitive organisms with DNA of the AT-type, which acquired photosynthesis upon entering symbioses with cyanobacterial organisms—that is, symbiogenesis.

Symbiosis on the Molecular Level: The Work of Ryzhkov

Symbiogenesis has been considered from a morphological or physiological position until now. The problem of unification in the evolution of heterogeneous bodies remains. The difficulty of the simultaneous unification of different genetic components entering symbiosis has been pointed out by several authors. In the Soviet literature, Vitalii Leonidovich Ryzhkov (1896–1977) tried to place this problem on the genetic level. Although his opinions seem naive (in the light of modern molecular biological data and the genetics of bacteria) and several seem simply mistaken (for example, his hypothesis regarding the nature of the recombinational processes of bacteria), in documenting the history of symbiogenesis, it is necessary to consider his opinions. Ryzhkov based his work on evidence accumulated in the 1960s; he simply could not know about data on the genetics of bacteria and viruses attained subsequently. From the perspective of the end of the 1970s, several of his generalizations seem unsupported—having the character of *Naturphilosophie*. But Ryzhkov's posing of the concept of symbiosis on the molecular level and his attempt, using evidence and theory, to reveal the evolutionary idea of the integration of organisms at the level of the unification of their genetic systems, are great contributions that justify detailed consideration of his work.

Ryzhkov's scientific interests involved three important branches of biology: cytoplasmic inheritance, viral diseases of plants, and concepts of theoretical evolutionary biology. I consider here only works directly related to symbiogenesis.

In the small pamphlet "The Rudiments of Biology: A History of Species and Individuals" (1924), Ryzhkov dwells on the concept of the nature of the mutual relations between organisms, emphasizing that the basis of their origin lies in the struggle for existence. In this struggle, organisms enter into various, more or less close, relationships. Sometimes this struggle leads "to enmity, in other cases—to mutual support" (Ryzhkov 1924, 155). The link between flowers that are pollinated by insects and the insects that pollinate them may serve as an example of a mutually beneficial relation firmly established in nature. Occasionally, the mutual linking of organisms of different species becomes more profound and close: these cases of a combined, mutually beneficial life may be called symbiosis.

Ryzhkov's published essays (1927) addressed problems then under intensive study, for example, cell evolution; the causes of changes in sex ratios; the chemistry of mutation and its causes, and mechanisms of evolution; he also addressed the problem of intracellular symbiosis. Intracellular symbiosis is the closer, mutually beneficial relationship between organisms in cases where one partner dwells within the cells of the other. The struggle for existence "compels organisms to come together variously; sometimes it imports one organism into the body of another. . . . In these or other relations, the organisms are beneficial to one another; then a combination is established between them called symbiosis. But if one organism turns out to be weak, then the other may take on a decisive predominance and the symbiosis becomes parasitism" (Ryzhkov 1927, 135–36).

Surveying the problem of cell evolution, Ryzhkov enumerated the proposed hypotheses, evaluating their scientific significance. He concluded that this problem is aided by the theory of symbiogenesis "better than by other theories," agreeing with its basic proposition that "various parts of the cell arose from different independent organisms, which at one time lived freely" (ibid., 26). Moreover, in considering the problem of the ancestral forms that initially entered into symbiosis, Ryzhkov denies that precellular organisms could have been similar to bacteria and cyanobacteria, as asserted in Merezhkovsky's concept of symbiogenesis.

In his attempt to theorize about the lability of genes in plants, Ryzhkov (1939) again turned for explanation to the study of symbiosis. While investigating the inheritance of mottled leaf patterns in tobacco, Ryzhkov described rare cases of labile genes determining leaf color (Ryzhkov 1927), a phenomenon studied in depth in corn. In several plants mottling was related to the synthesis of anthocyanin in the aleurone layer. Phenotypically among those forms that regularly developed anthocyanin in the endosperm and those that did not form anthocyanin were also forms in which the synthesis was labile: the endosperm had some spots with anthocyanin and others lacked it. To Ryzhkov, the unique thing about this phenomenon was that there were genes that influenced other parts of the chromosome that could, in many cases, move from one chromosome to another. This allows one to speak of "wandering" or "jumping" genes [mobile genetic elements], unparalleled in genetics. An explanation of genetic control in cases of paramutation, as Ryzhkov asserted, was related to the wandering agents. Ryzhkov supported the proposition that wandering agents were a kind of virus living symbiotically in plant cells and resembling latent phages, or, as R. Brink thought, similar to episomes of bacteria (Ryzhkov 1965). Ryzhkov extensively defined the concept of wandering genes much later. He meant extrachromosomal genetic entities in cells capable of infecting other cells that lack them; they are now known as plasmids and episomes (Ryzhkov 1976, 93).

A central direction of Ryzhkov's scientific activity was the study of viral diseases in plants. The results of many years of work on biochemistry and physiology of viral diseases are found in the monograph *The Foundations of the Study of Viral Diseases in Plants* (1944), awarded the State Prize of the USSR in 1946. Together with this research he completed a cycle of work directed at the explanation of the nature and behavior of viruses. As early as 1938 Ryzhkov concluded that viruses in plants act as genuine parasites. Viruses, like other parasitic organisms, possess adaptations that secure for them an existence as a source of infection. These simple mechanisms and subtle cellular parasites, completely dependent on their hosts, manifest simultaneously a distinct autonomy and capacity for evolution (Ryzhkov 1942).

These opinions, stemming from extensive evidence from his study on viruses and generalized from a wider biological point of view, lay at the base of Ryzh-kov's concept of symbiosis at the molecular level. He made his most complete presentation of the evolutionary significance of symbiosis at the molecular level in a series of articles (1960–76).

Virus study revealed new problems of intracellular symbiosis, but this time at the molecular level. Ryzhkov thought that symbiosis at the molecular level included, to various degrees, mutually beneficial cases of the combined life of several bacteria, protists, insects, and plants with forms resembling the largest molecules. The analysis of this phenomenon turned out to be fraught with greater difficulties than those that arose during the direct study of intracellular sym-biosis, since in the presence of a way to study macromolecular cell components, it is remarkably difficult to distinguish the cell components themselves from ex-ogenous macromolecules that penetrate the cell and reproduce inside it. At this level, it is as if the fine border between cell organelles and symbiotic organisms that entered the cell from outside were erased. Studies of integration at the molecular level, therefore, could be useful for proving the possibility of the origin of cell organelles by symbiosis.

Ryzhkov introduced data for the hypothesis of symbiosis on the molecular level. He considered lysogeny to provide the most convincing evidence. The temperate phage penetrates and enters into close contact with the bacterial cell. The union of the temperate phage and the cell-host can be accompanied by a profound integration of their genetic systems into a single whole, in which the genetic material of the phage is integrated into the net of the bacterial DNA and reproduces in the status of a prophage as a part of it. Moreover, prophages preserve a distinct degree of autonomy and, in a certain sense, live their own lives. The establishment of a relation between the bacterial cell and the temperate, lysogenic phage leads to an alteration of the genetic attributes of the bacterium.

In Ryzhkov's opinion, the insertion of the genome of the phage into the bacterial genophore was convincing evidence of the symbiotic integration of the genetic apparatus of the partners entering into a mutual existence, a single system. He considered the union of distinct strains of *Escherichia coli* with the phage *lambda*, dependent on the ability of the bacterium to ferment lactose, an example of a similar symbiosis. Another example was the formation of toxic strains of diphtheria bacteria after the contact of nontoxic strains with the dying diphtherial phage. L. A. Zil'ber (1952) supported this opinion regarding the symbiosis of several viruses and bacteria. L. Hewitt (1956) also wrote about the fact that the treatment of the bacterial cell by the phage can lead to the arising of "a symbiotic infection." Ryzhkov's opinion that lysogeny is an instance of sym-biosis may be farfetched, since a phage is always a potential enemy, a parasite, and upon the induction of a prophage, the cell dies. But Ryzhkov wrote that lysogeny, at any given moment, is a good example of symbiosis. The level of utility of the relation formed between the bacterium and the temperate phage allows one to speak of genuine symbiosis in cases of lysogeny. The production of a toxin that leads to the death of host tissue is useful to the diphtheria bacterium. Moreover, lysogenic bacteria lyse to some extent and in doing so release phages that attack sensitive bacteria. Ryzhkov wrote, "Perhaps this is beneficial to the lysogenic population, since it facilitates the elimination of other, nonlysogenic, strains of bacteria" (1965, 386).

Support for the hypothesis of symbiosis at the molecular level comes from phenomena such as episomes in bacteria—genetic elements that are found both in a free state and in integration with the genophore. These episomes (including the sex factor F) control a function connected with bacterial conjugation, ensuring the resistance of bacteria to a series of antibiotics, as well as episomes that induce the synthesis of colicin proteins.

Ryzhkov used the term *episomes* in the broad sense commonly accepted in the 1950s and 1960s. He meant plasmids [small replicons, determinants of heredity present in the cytoplasm of bacteria that replicate autonomously relative to the genophore], as well as genuine episomes [bacterial plasmids capable of incorporating themselves into the bacterial genophore and becoming part of its large genome]. Here the term is used as Ryzhkov did.

Ryzhkov felt that the origin of episomes was obscure. Based on similarities between dying phages, which he considered to be intracellular symbionts, and episomes, he hypothesized about their symbiotic nature.

Ryzhkov (1965, 387) thought the ability to infect and autonomously replicate to be the most important characteristic linking episomes to temperate phages. Episome similarity to the temperate phage was particularly clearly observed in colicinogenic episomes and F, the fertility factor (Ryzhkov 1960, 516). Sex factor F is acquired by bacterial cells by infection during conjugation and can be removed by chemical treatment. Like a phage, the F factor can autonomously replicate. Interaction occurs between the sexual factor and other episomes of bacteria: for example, if the RTF factor is present in the donor together with F+, the sexual factor, transfer from the donor is neutralized; yet when RTF is in the recipient, F-factor passage is also neutralized, though to a lesser extent (Watanabe and Fukasawa 1962). Thus, the episome (factor F) that controls the sexual process in bacteria is a phagelike life form and "in this case, we are talking about symbiosis at the molecular level" (Ryzhkov 1965, 388). Hayes (1965, 476), who held a similar view on the similarities between the sex factor and temperate phages, a prototype of which is *lambda* phage, rejected speculation about its origin.

Studies of the killer factor in paramecia and the kappa factor are valuable to Ryzhkov not only to provide a molecular basis for symbiosis but for symbiogenesis concepts generally. He was impressed that the presence of kappa depends on the nuclear genetic constitution of the paramecia, attesting to profound integration of the symbionts, whose study is already at the limit of research capabilities.

Symbiosis at the molecular level is also illustrated by the wide dissemination of viruses of arthropods, although, as Ryzhkov emphasized, few cases have been well studied. Increased sensitivity to CO_2, as well as the reproductive isolation of several lines of *Drosophila,* provide telling examples of symbiotic integration and inheritance of an external agent with insect cells. Here one encounters the interesting difficulty of differentiating between genuine nuclear inheritance and cytoplasmic infection, or "infective" inheritance. In several strains of fruit fly an increased sensitivity to CO_2 was discovered (Teissier and L'Héritier 1937), and L'Héritier (1958) showed that this trait was transmitted to offspring in the ovule cytoplasm of the lineage of the original female. A reduction in the stability of the fruit fly in the presence of CO_2 was expressed by infection via the cytoplasmic factor *sigma:* factor sigma is easily transmitted by inoculation of the ooplasm from a carrier to a fly free of it (the sigma factor), and after such transfer it is transmitted to the offspring. Sigma factor is a large virus, as determined by filtration. Subsequently, the attributes of viral mutants distinguished by the ease with which they infect the fruit fly were studied. Sensitivity to CO_2 in *Drosophila* must be viewed as "an irrefutable case of infection, . . . though for now it remains a mystery whether this is a case of symbiotic infection or whether we are talking about parasitism" (Ryzhkov 1965, 393).

Reproductive isolation in *Drosophila* was seen by the discovery in Brazil of a factor that leads to the death of male zygotes, a consequence of which is that the females carrying this factor produce only female offspring (Malogolovkina and Paulson 1957). This infective "sex relation" factor can be transmitted by normal females by injection of the ooplasm. This factor is a small spirochete [later shown to be a spiroplasma] (Paulson 1961).

Ryzhkov concluded that all of these cases provide examples of infective heredity, demonstrating symbiosis on the molecular level and incontestably proving the significance of symbiosis as a source of new inherited characteristics. These macromolecular agents, autonomous, self-reproducing forms with which others can be artificially infected, are facultative elements of the cell.

Ryzhkov connected symbiosis at the molecular level with its significance in evolution. Although he rejected a universal significance of symbiosis in evolution, symbiotic integration of the host cell genome with exogenous macromolecular life forms was evident to him. Symbiosis on the molecular level appeared in the form of latent phages and episomes in bacteria, or infective heredity in *Drosophila* as factors that changed the cell's genetic potential. Ryzhkov's work on symbiogenesis at the molecular level was not limited to this; he also hypothesized about recombination in bacteria (Ryzhkov 1960).

He claimed that genetic recombination in bacteria—conjugation, transformation, and transduction—do not affect the bacterial genome. This process is based on genetic exchange between the phagelike genetic structures in symbiosis with the host cell (Ryzhkov 1960, 519).

The development of bacterial genetics and molecular biology saw Ryzhkov's supposition abandoned. Chromonemal recombination was proved (Jacob and Vollman 1962; Hayes 1965; *Elementary Genetic Processes* 1973). In the quickly moving world of modern science, opinion changes by deeper study; nevertheless, Ryzhkov's hypothesis should be remembered for its unique explanation of the mechanism of prokaryotic recombination. Ryzhkov considered his hypotheses significant in attracting current attention to concepts of symbiosis and symbiogenesis, particularly at the molecular level. His ideas about the symbiotic union of the bacterial genome with that of temperate phages and episomes and about the symbiotic basis of bacteria recombination, leading to the alteration of inherited traits (the most important precondition of the evolutionary process), well expressed the evolutionary essence of symbiosis, particularly with respect to the integration of genetic systems.

Although emphasizing the evolutionary significance of symbiosis at the molecular level as a source of change in inherited characteristics in bacteria, Ryzhkov admitted this phenomenon to be a factor in the evolutionary origin of certain eukaryotic cell organelles. He accepted the evolutionary origin of organelles with great reservation and particularly criticized the ideas of Kozo-Polyansky, who in Ryzhkov's opinion illegitimately developed the hypothesis of the general significance of symbiogenesis.

Ryzhkov generalized the data, showing that the intracellular symbiosis of host cells with temperate phages and episomes can be a means to acquire new genetic attributes, one increasing their genetic continuity; thus Ryzhkov's ideas are interesting for the history of symbiogenesis.

Symbiogenesis and the Origin of the Eukaryotic Cell: A General Phylogeny of Life

I anticipate consideration of these problems by quoting the remarkable Russian scientist M. S. Tswett, whose words even today precisely encompass the entire problem: "The hypothesis of the symbiotic origin of the cell, no matter how improbable it seems, contains, however, nothing untenable" (Tswett 1896, 573).

The modern data, showing the abilities of plastids and mitochondria and revealing their profound similarity with certain free-living bacteria, have been briefly reviewed. Here we consider the modern interpretation of the question of the origin and evolution of eukaryotic cell organization.

The evolutionary transformation of the prokaryotic cell into the eukaryotic [cell], a problem raised by many biologists, is again on the agenda, thanks to an accumulation of relevant data on cell organelles. The fundamental question is: How did the transformation from prokaryotes to eukaryotes occur? Did the organelles of the eukaryotic cell arise by gradual differentiation, or is it more correct to relate their origin to symbiosis? The first hypothesis is more difficult to reconcile with new facts, whereas the symbiotic model is well supported. The large quantity of new data serves to justify the serious reconsideration of symbiogenesis in cell evolution (Margulis 1970, 11). An atmosphere similar to that of the middle of the nineteenth century following the confirmation of the cellular theory has currently developed. The Soviet botanist S. S. Khokhlov correctly remarks: "Under the pressure of new facts and ideas, the doctrine of the cell changed. And, very likely, precisely here the most revolutionary shifts are going on, which are breaking old hypotheses and confirming new general biological principles. One could say that we are witnesses to changes in the doctrine of the cell that could be justly called the second discovery of the cell, in which the veil is torn from the historical process of the origin of the cell itself and in which the laws of the structure and functioning of the separate organelles and of the cell as a whole are seen in a new light. All of this, in its turn, decisively compels us to reconsider not only the general theoretical position of biology, but of the applied sciences and medicine as well" (Khokhlov 1977, 6).

The hypothesis of the origin of the eukaryotic cell by symbiosis was revived in modern form by the American researcher L. Sagan-Margulis (Sagan 1967; Margulis 1970, 1971a, 1971b, 1974, 1975) and was strongly supported by a series of authors (Kirk 1970; Roodyn and Wilkie 1970; Takhtadzhyan 1973; Rutten 1973; Yablokov and Yusufov 1976; and many others).

Margulis's position is that two basic facts cannot be explained by the traditional hypothesis of the origin of eukaryotes from a single ancestral population of prokaryotes by accumulation of small, step-by-step mutations. The first is a clear discontinuity between prokaryotes and eukaryotes (Margulis 1970, 10–12). The absence of intermediate links between the algae and the blue-greens—from which the former take their origin as, for example, Fritsch (1935) and Bessey (1950) suggested (Margulis 1970)—is confirmed by all modern cytological, geological, and biochemical data. The second unexplained fact is the absence of mitosis in most microbes; in the bacteria and cyanobacteria, there is no mitosis homologous to that of plants and animals (ibid., 22).

Moreover, Margulis asserts, the hypothesis regarding the origin of mitochondria, plastids, and the [kinetosomes] basal granules of flagella and cilia by symbiosis corresponds completely with the features of these organelles, if one accepts that they arose as endosymbionts. The criteria that allow unmistakable distinction between organelles of endosymbiotic origin and those of intracellular differentiation can be reduced, according to Margulis, to the following characteristics. Symbiogenetic organelles must have their own DNA synthesized and regulated within them, and they must contain an endogenous protein-synthesizing apparatus. They must be able to replicate, by means of which offspring cells receive a copy of their genome. They have no intermediate intracellular stages: the loss by the cell of the symbiotic organelles must be accompanied by the loss of all metabolic properties that depend on their genome. The inheritance pattern of organelles need not correspond to the rules of Mendelian genetics.

According to Margulis, the first eukaryotes arose in the late Precambrian eon from prokaryotic, free-living ancestors through a series of specific symbioses. The major steps in the evolution of living organisms are: the initial ancestral forms were fermentative, heterotrophic prokaryotes capable of the anaerobic decomposition of abiotic, organic substances. In the early Precambrian eon, under the uninterrupted pressure of natural selection, an extensive adaptive expansion of the prokaryote's metabolic activity arose.* Several types of these forms evolved. First were mycoplasm-like, fermentative heterotrophs capable of anaerobic chemoautotrophy (chemosynthesis), allowing them to feed in the absence of abiotic, organic compounds. Then came spirochetes and photosynthesizing prokaryotes, including those of the coccoidal blue-green type, as well as aerobic Gram-negative bacteria capable of Kreb's cycle metabolism (promitochondria).

The constant saturation of Earth's atmosphere by oxygen during the period beginning about 1,200 million years ago, as a result of the activity of cyanobacteria, was that to which the constant action of natural selection must have inevitably adapted prokaryotes. To survive conditions that increasingly inhibited the formation of abiotic, organic compounds, the mycoplasm-like heterotrophs, having previously acquired nutrients by anaerobic decomposition, must have absorbed organic substances formed by photo- and chemosynthesis. The evolution of feeding methods led to a stage at which several types were capable of absorbing smaller bacteria (promitochondria) without killing them. The capture and penetration of tiny, aerobic bacteria, their establishment within the cell of mycoplasm-like bacteria, and the further stabilization of similar endosymbiotic

---

*A detailed consideration of the concrete, metabolic paths of the transition from anaerobiosis to aerobiosis and of the possible mechanisms of connection between these biochemical processes and of the morphological evolution of cellular organization has been presented by G. B. Gokhlerner (1977).

relations led to the formation of large amoeboids that maintained their mitochondria. The formation of similar symbiotic cell structures with profound metabolic intensity was supported by selection.

The change in cell metabolism created the necessity of intracellular differentiation and, most importantly, of the formation of a nuclear membrane, which was, Margulis suggests, a unique barrier for defending the nucleocytoplasmic DNA from the surrounding cytoplasmic mitochondria, with its oxidizing enzymes.

The second step in eukaryote evolution occurred when amoeboids that symbiotically maintained mitochondria acquired highly motile, spirochete-like bacteria.[†] This led to the formation of heterotrophic, spirochete-like, undulipodiated forms. The selective value of the acquisition of kinetosomes and their undulipodia was enormous. A most important adaptation, it significantly increased the locomotive possibilities of the amoeboids. Moreover, the exceptional biological usefulness of such a symbiosis, according to Margulis, lies in the fact that the spirochete-like bacteria gradually took on another important role: they enabled the formation of a genuine nucleus with typical, mitotic division.

The spirochete-like symbionts initiated the microtubular structure of eukaryotic organisms, an organization of the type (9+2), according to Margulis. The basal body [kinetosome], having grown the rest of the undulipodium, was similar in ultrastructure to centrioles (type 9+0); it began to take part in cell division, which, through a series of morphological processes, led to the formation of mitotic centrioles, spindle and chromosomal centromeres. In *The Origin of Eukaryotic Cells* (1970), Margulis analyzes micrographs of mitosis in a series of eukaryotic microorganisms and on this basis tries to distinguish stages in the evolution of mitosis. Evaluating this attempt, Takhtadzhyan noted that the images observed in early eukaryotes would be better considered as various models of the mitotic apparatus that reflect nature's attempts to create the most perfect mitotic mechanism. If, in the first stages of cell evolution, the basal body itself functioned as the mitotic center, then it would later "have served only as the genetic matrix for the synthesis of microtubular structures homologous to it, including the centromeres of the chromosome" (Takhtadzhyan 1973, 25). In the process of further evolution, heterotrophic, amoeboid, and mastigote forms, the simplest eukaryotic organisms of the type of several modern "rhizomastigotes" such as *Chaetoproteus* (Mastigamoeba), formed the beginning of the animal kingdom. From colorless amoeboids arose the unique group of heterotrophs, the fungi.

The final (third) symbiotic step in the evolution of eukaryotes included the

---

[†Eds. note—Probably the first symbiosis has acquisition of undulipodia. All of this material has been revised; see Margulis 1992.]

union of various populations of protoeukaryotic heterotrophic amoebomastigotes with photosynthesizing prokaryotes of coccoidal cyanobacteria types. The evolutionary significance of this symbiosis was great, since it made possible the development of autotrophic nutrition. The origin of autotrophs altered the entire system of trophic relations of the biosphere, increasing its productivity. Entering into the cell of the amoebomastigote, the prokaryotes, thanks to the establishment of stable, hereditary symbiotic relations, became obligate-symbiotic, photosynthesizing plastids. This began the development of the plant kingdom. In general this is the symbiotic model of the origin and evolution of the eukaryotic cell, as proposed and developed by Margulis. The theoretical significance of this hypothesis is that it satisfactorily explains the origin of the multigenomic nature of the eukaryotic cell. It naturally follows that all eukaryotes are at least bigenomic organisms with cells that contain two independently arising genomes: the original prokaryotic host organism and that of the mitochondria. The cells of the majority of heterotrophic eukaryotes are trigenomic forms, whereas most photosynthesizing eukaryotes are quadrigenomic forms (Margulis 1975, 23).

This hypothesis of the origin of eukaryotic organization is actively discussed and receives support from a number of Russian and foreign scientists. The proposition of the evolution of the cell corresponds well with contemporary cytological, biochemical, and paleontological data, explaining the nature of cell organelles.

Hypotheses that interpret eukaryotic cell evolution without symbiosis have also received the support of many biologists. Ignoring the differences in the approaches to a nonsymbiotic explanation of the origin and evolution of cell organelles, all of these hypotheses acknowledge the formation de novo of these structures in phylogenesis by progressive intracellular differentiation of prokaryotes. The strictest among them, laid out in a precise form by Taylor (1976), is the hypothesis of the complete autogenous origin of eukaryotes.

The autogenous hypothesis has advantages over serial endosymbiosis theory (SET, using the author's terminology, although the latter, as Taylor acknowledges, corresponds best to new data on the ultrastructure of organelles. The main advantages of the autogenous hypothesis is that it is the most simple; it does not demand the establishment of compatibility mechanisms for the organelles and their environment, such as the abolition of organelles by digestion or ejection, their synchronous reproduction, and the integration of host-organelle metabolism. Moreover, the autogenous hypothesis explains why nuclear genes control many internal organelle processes and "why the DNA of the organelles, particularly the mitochondrial DNA, resembles the plasmid DNA to a greater degree than the original genophore of the prokaryotes" (Taylor 1976, 386).

Eukaryotes, according to the autogenous hypothesis, evolved directly from prokaryotes, developing from a cyanophyte-like, photosynthetic ancestor, the "*Uralga.*" The plastids and mitochondria, structural elements of the cell, were formed by progressive differentiation of the membrane system of prokaryotes, leading to the isolation (compartmentalization) of specialized, metabolic components. From this scheme of development, a series of consecutive stages in the transformation of the prokaryotic cell can be constructed. This series begins from an initial form, a prokaryote with a diffused genome and multipurpose membranes, a cytoplasmic matrix with 70S ribosomes, and polar granules represented by hypothetical primitive, microtubular centers. The second stage involves the formation of a large nucleoid, which arose from the expansion of one of the diffused genophores. The third stage includes the origin of a nuclear membrane from thylakoid membranes. The gradual differentiation of several membranes that fulfill specialized functions, for example, photosynthesis and respiration, occurs in the fourth stage. The general membrane system alterations and the union of microfibrils and microtubules occur in the fifth stage, the phase of absorption. The process of absorption is related to the ability of membranes to move dynamically (cytosis), thanks to which occurred the isolation of several of the peripheral genomes and the large and small segments of the surrounding membrane specialized for photosynthesis and respiration. The final stages—six, seven, and eight—are related to further differentiation of the original cell matrix and its membranes and with the formation of a microstructure (type 9+2) and flagella.

The hypothesis of the autogenous origin of the eukaryotic cell fundamentally supposes complex transformation of the membrane systems of the blue-green algal prokaryotes and their ability to compartmentalize the cell cytoplasm. This is particularly well observed in the organization of the membrane system in several blue-green algae. A tendency toward the formation of partially exclusive spheres of cytoplasm at the expense of interconnections of thylakoid membranes has been noted in Nostocales (Gromov 1976). The possibility for other methods of compartmentalization was established in the single-celled alga *Synechococcus* sp. The thylakoid membranes unite with their ends to the plasma membrane, as a result of which the limited spheres of cytoplasm contain two types of membrane. "If one bears in mind the well-known significance of the cytoplasmic membrane for the replication of the DNA, compartmentalization of this type, perhaps, must be considered as having great evolutionary significance" (Gromov 1976, 78). The characteristic organization of the blue-green alga membrane system may merely reflect the individual properties of the separate forms. They undoubtedly testify, however, to "distinct tendencies in the development of the membrane systems of the prokaryote" (ibid.). In evaluating the possible evolutionary significance of compartmentalization in the origin of cell structure, Gromov concludes that the origin of membrane-bounded organelles by compartmentalization must be considered no more speculative than any other proposition.

The hypothesis of the autogenous (or *direct*) origin of eukaryotes, in attempting to explain this phenomenon, contradicts the symbiosis hypothesis. It relies importantly on the progressive development of the prokaryotic membrane system and requires, to a still greater extent than symbiogenesis, confirmation by cytological and particularly ultrastructural data. The most serious difficulty for all of the nonsymbiotic hypotheses of eukaryote origin is the absence of forms intermediate between prokaryotes (for example, blue-greens) and genuine eukaryotic organisms. Organisms serving as a basis for transitional stages designated in the autogenous model still have not been observed in nature. The existence of intermediate forms has not been confirmed, even after careful investigation of the ultrastructure of cyanobacteria (Gromov 1976). Another vulnerable spot in the autogenous origin of eukaryotes is the explanation of a true genome—determining genetic and biochemical dynamic autonomy—in plastids and mitochondria, as noted by Zhukova (1975).

The construction of an organized system of diversity in the living world depends on elucidation of the relations between systematic groups, including higher taxa. The construction of systematics adequately reflecting evolution and the origin of higher taxa has serious objective difficulties, the most important of which is the absence of experimentally elucidated phylogenetic relations. Systematics of higher taxa is essentially a theoretical science, and that is why it is so unpopular among empirically inclined scientists, as Kozo-Polyansky (1948) emphasized. A scientific approach to the phylogenetic relations of individuals and the origin of higher taxa is possible, he nevertheless insisted. This approach requires the profound synthesis of data from all branches of biology and a knowledge of the forces and laws of evolution.

A solution to the origin of eukaryotic organisms from prokaryotic ancestors is extremely significant for the construction of a general phylogeny. This problem is closely related to the evolutionary origin of cell organelles, each with its relative genetic and biochemical independence and the ability to reproduce (Stanier 1970). Symbiogenesis in eukaryotic cell origins provides a persuasive explanation. Interest in the symbiogenesis hypothesis has grown from working out the higher taxa (Margulis 1968, 1974, 1975; Raven 1970; Lee 1972; Takhtadzhyan 1973, 1974, 1976).

Intense attention to symbiosis in evolutionary relationships is also linked to the analysis of macroevolution. Natural selection of subtle, undirected mutations in individuals is considered demonstrated as a leading factor of evolution at the species level (microevolution). The following questions about macroevolution are appropriate: Is this a mechanism of defining all evolutionary transformations in higher taxa? Do specific factors in the origin of large-scale systematic [higher taxa] differences exist? Abandoning all teleological hypotheses of macroevolution, we still must pay attention to the role of symbiosis as a mechanism in phyletic evolution (Zavadsky 1968).

Interest in symbiosis as a factor in the evolution of higher taxa leads to a reconsideration of hypotheses regarding the general systematics of the living world. Dictated by the level of modern biological knowledge, new data reveal the inadequacy of the traditional division of the living world into two kingdoms, plants and animals (Whittaker 1969; Margulis 1971b; Takhtadzhyan 1973, 1974).

The acquisition of previously absent information concerning cell organization in various organisms shows two unique and sharply isolated groups in nature: the prokaryotes (prenuclear) and eukaryotes (nuclear). Bacteria, including cyanobacteria, belong to the first group, and all remaining organisms to the second. The differences between prokaryotes and eukaryotes, as has gradually been revealed, are much more profound and fundamental than, for example, those between animals and plants (Takhtadzhyan 1974, 54). The most complete characterization of the differences between prokaryotes and eukaryotes is presented in the works of L. Margulis.

The division of the living world into prokaryotes and eukaryotes and the recognition of the possibility of a symbiotic origin of eukaryotic cells are reflected in phylogenetic macrosystems. The most thoroughly developed of these is the four-kingdom system of Takhtadzhyan (1973), in which the entire living world is divided into two large groups of organisms, each of which is accorded the rank of superkingdom. These are the superkingdoms of prenuclear organisms (Prokaryota) and the superkingdom of nuclear organisms (Eukaryota).

The superkingdom of prokaryotes includes only one kingdom—Schizophytes (Mychota), within which are two subkingdoms: bacteria (Bacteriobionta) and blue-greens (Cyanobionta). In the superkingdom of eukaryotes are three kingdoms: animals (Animalia), fungi (Mycetalia), and plants (Vegetabilia). In the animal kingdom is the subkingdom protozoa (Protozoobionta) and the subkingdom multicellular animals (Metazoobionta). The kingdom of fungi includes the lower fungi (Myxobionta) and higher fungi (Mycobionta) subkingdoms. The plant kingdom includes red algae (Rhodobionta), true algae (Phycobionta), and higher plants (Embryobionta) subkingdoms. Takhtadzhyan's system of organisms differs fundamentally from earlier classifications. Grounded in evolution, it reflects the evolutionary development of higher taxa. Thus the concept of symbiogenesis is productive for understanding the evolutionary origin of higher taxa in the living world.

CONCLUSION

The history of the concept of symbiogenesis and discussion about it has been documented here. The historical and critical approach of this book compels us to detail the history of the concept; its current status was discussed only briefly.

I have concluded that the research of Russian scientists played a leading role in developing the concept of symbiogenesis, particularly in the early stages. These early stages were therefore considered at the greatest length.

The concept of symbiogenesis clearly must be considered deserving of further scientific elaboration. In this century it continues to stimulate investigative thought in various branches of biology. One can justly consider it a working hypothesis.

Symbiogenesis is a firmly established fact with regard to the evolution of lichens, mycorrhizae, rhizobia in the roots of leguminous plants, and symbioorgans in many species of plants and animals. Symbiogenesis qualifies with regard to these examples as a fully developed scientific theory.

Symbiogenesis is also possible as an explanation of the evolution of basic organelles of the eukaryotic cell. Attempts to elevate the concept of symbiogenesis prematurely to the rank of a universal and general biological theory should proceed with significantly more caution.

The material analyzed here, particularly that presented in chapter 7, shows that symbiogenesis rests on a significantly more broad factual basis than it did half a century ago. Moreover, the foundation differs qualitatively in that it includes molecular biological and cytological data. The contemporary status of symbiogenesis requires acknowledgment of symbiosis as a possible factor in the evolution of such major lineages as eukaryotes.

Symbiosis is emerging as a mechanism in the evolution of several groups of prokaryotes and eukaryotic single-celled and multicellular organisms. Symbiosis is emerging, one may say, in the world of prokaryotes as a major mechanism of evolution.

Symbiogenesis is emerging, too, as one of the natural regularities of phylogeny. A series of consecutive stages of the integration of organisms as symbionts has been retraced from the community level to the physiological level, that is, the formation of new combinations of individuals. The analysis of the concept of symbiogenesis compels one to consider the need for an addition to the principle of divergence: (evolution by the isolation of groups) the principle of symbiogenesis [anastomosis of branches on phylogenetic trees].

The idea that complexification of organization in evolution is based on prolonged symbiosis between phylogenetically diverse organisms has become extremely productive. With the acceptance of symbiogenesis, serious difficulties that impeded explanation through the evolutionary theory of complexification in phylogenesis fall away.

One hopes that the significance of symbiosis will soon be evaluated so that its role as a mechanism in evolution will occupy an appropriate place in contemporary evolutionary theory.

REFERENCES

Afanas'ev, V. A. 1937. Parasitism and symbiosis. In *Problems of General Parasitology*. Leningrad-Moscow, 15–20. (In Russian.)

Ahmadjian, V. 1971. The lichen symbiosis: Its origin and evolution. In *Evolutionary Biology*, vol. 4. New York.

Aleksandrov, V. G. 1950. The question of the origin of green plastids in plant cells. *Botanical Journal* 35(5): 475–81. (In Russian.)

Antonov, A. S. 1969. DNA structure and the systematic position of organisms. *Advances in Contemporary Biology* 68(3): 299–317. (In Russian.)

———. 1974. Genosystematics: Achievements, problems, and perspectives. *Advances in Contemporary Biology* 77(2): 31–47. (In Russian.)

Ball, G. 1969. Organisms living on and in protozoa. In *Research in Protozoology*, no. 3.

Baltus, R., and I. Brachet. 1963. Presence of deoxyribonucleic acid in the chloroplast of *Acetabularia mediterranea*. *Biochimica et Biophysica Acta* 76:490–92.

Baranetsky, O. 1868. Independently living gonidia of lichens. *Transactions of the First Russian Congress of Naturalists and Physicians, Botanical Section*. St. Petersburg, 45–59. (In Russian.)

Barricelly, N. A. 1963. Numerical testing of evolution theories. Part 1, Theoretical introduction and basis tests. *Acta Biotheoretica* 16(1–2): 70–98.

———. 1963. Numerical testing of evolution theories. Part 2, Symbiogenesis and terrestrial life. *Acta Biotheoretica* 16(3–4): 99–126.

Bary, A. de. 1879. *Die Erscheinung der Symbiose*. Strassburg.

Batra, S. W. T., and L. R. Batra. 1967. The fungus gardens of insects. *Scientific American* 217(5): 112–20.

Bazilevskaya, N. A. 1959. Creative activities of B. M. Kozo-Polyansky. *Transactions of the Institute of the History of Natural Sciences and Technics* 23(4): 324–57. (In Russian.)

Bazilevskaya, N. A., I. P. Belokon', and A. A. Shcherbakova. 1968. *A Brief History of Botany*. Moscow. (In Russian.)

Beale, G. H., A. Jurand, and J. R. Preer. 1969. The classes of endosymbionts of *Paramecium aurelia*. *Journal of Cell Science* 5:65–78.

Beklemishev, V. N. 1964. *Basic comparative anatomy of invertebrates*, vol. 1. Moscow. (In Russian.)

———. 1970a. Organism and association. In *The Biocoenotic Bases of Comparative Parasitology*. Moscow, 26–42. (In Russian.)

———. 1970b. Parasitism of the limbs of terrestrial vertebrates. In *The Biocoenotic Bases of Comparative Parasitology*. Moscow, 250–60. (In Russian.)

References

Belozersky, A. N. 1969. *Nucleic Acids and Their Connection with Evolution, Phylogenetics, and the Systematics of Organisms.* Tashkent. (In Russian.)

Belozersky, A. N., A. S. Antonov, and B. M. Mednikov. 1972. Introductory article, *DNA Structure and the Systematic Position of Organisms.* Moscow, 3–16. (In Russian.)

Belozersky, A. N., and B. M. Mednikov. 1972. *Nucleic Acids and the Systematics of Organisms.* Moscow. (In Russian.)

Beneden, P. van. 1876. *Die Schmarotzer des Thierreichs.* Leipzig.

Berg, L. S. 1922. *The Theory of Evolution.* Petrograd. (In Russian.)

Beridze, T. G., and M. S. Odinstova. 1969. Deoxyribonucleic acid of cytoplasmic structures: Plastids and mitochondria. In *Successes of Biological Chemistry,* vol. 10. Moscow, 36–63. (In Russian.)

Bernard, N. 1909. L'evolution dans la symbiose: Les Orchidées et leurs champignons commensaux. *Ann. sci. natur.* 9(1): 1–196.

Biographical Dictionary of Professors and Teachers at St. Petersburg University, vol. 2. 1898. St. Petersburg, 33–36. (In Russian.)

Blagoveshchensky, A. V. 1957. B. M. Kozo-Polyansky. *Bulletin of the USSR Botanical Garden* 28:123–24. (In Russian.)

Blyakher, L. Ya. 1971. *The Problem of* the *Inheritance of Acquired Characteristics.* Moscow. (In Russian.)

Bonen, L., and W. F. Doolittle. 1975. On the prokaryotic nature of red algal chloroplasts. *Proceedings of the National Academy of Sciences* (USA) 72:2310–14.

Borodin, P. 1919. Andrei Sergeevich Famintsyn (1835–1918). Obituary, *Journal of the Russian Botanical Society* 4(1): 132–51. (In Russian.)

Breslavets, L. P. 1959. The history of questions on the origin of chloroplasts. *Transactions of the Institute of History of Natural Science and Technology,* vol. 23. *History of Biological Science* 4:257–88. (In Russian.)

———. 1963. Contemporary ideas on the origin of plastids. *Proceedings of the Soviet Academy of Sciences, Biological Series* 1:91–98. (In Russian.)

Buchner, P. 1965. *Endosymbiosis of Animals with Plant Microorganisms.* Wiley Interscience, New York.

Butenko, R. G. 1971. *From Free-Living Cells to Plants.* Moscow. (In Russian.)

Cheng, T. (ed.). 1971. *Aspects of the Biology of Symbiosis.* London.

Cheremisinov, N. A. 1959. Session of the Voronezhian division of the Soviet Botanical Society, honoring the birth of B. M. Kozo-Polyansky. *Botanical Journal* 44(2): 275. (In Russian.)

———. 1962. Kozo-Polyansky as a teacher and scientist-phytopathologer. *Proceedings of the USSR Academy of Sciences, Biological Series* 2:275–82. (In Russian.)

Darlington, K. 1972. *Evolutionary Botany.* Moscow. (In Russian.)

Danilov, A. N. 1910. The interrelationships between the gonidia and the fungal component of the lichen symbiosis. *Proceedings of the St. Petersburg Botanical Garden* 10(2): 33–70. (In Russian.)

———. 1921. Symbiosis as a factor in evolution. *Proceedings of the Main Botanical Garden of the Russian Soviet Federated Socialist Republic* 20(2): 122–36. (In Russian.)

References

————. 1927. *Nostoc* in symbiosis. *Russian Archive of Protistology* 6(1–4): 83–92. (In Russian.)

————. 1929. Introduction to synthesis of the lichen *Leptogium issatschenkoi* Elenk. *Proceedings of the Main Botanical Garden of the USSR* 28(3–4): 225–64. (In Russian.)

————. 1933. Lichen symbiosis. *Nature* 11:34–44. (In Russian.)

Darwin, C. 1939. Origin of species. In *Works*, vol. 3. Moscow, Leningrad, 253–680. (In Russian.)

Dazho, R. 1975. *Basic Ecology*. Moscow. (In Russian.)

Dogel', V. A. 1951. *General Protistology*. Moscow. (In Russian.)

————. 1962. *General Parisitology*. Leningrad. (In Russian.)

Dubinin, N. P. 1976. *General Genetics*. Moscow. (In Russian.)

Dubos, R., and A. Kessler. 1963. Integrative and disintegrative factors in symbiotic associations. In *Symbiotic Associations*. London.

Elementary Genetic Processes. 1973. Leningrad. (In Russian.)

PUBLICATIONS BY A. A. ELENKIN

1901. Facultative lichens. Minutes of sessions, *Transactions of the St. Petersburg Society of Natural Science* 32(6): 261–69. Pt. 1 of a series. (In Russian.)

1902. The question of "internal saprophytism" ("endosaprophytism") in lichens. *Proceedings of the St. Petersburg Botanical Garden* 2(3): 65–84. (In Russian.)

1904. New observations of the phenomenon of endosaprophytism in lichens. *Proceedings of the St. Petersburg Botanical Garden* 4(2): 25–39. (In Russian.)

1907a. The concern of lichen symbiosis for the evolution of organisms. *Transactions of the St. Petersburg Society of Natural Science* 38(4): 160–75. Pt. 1 of a series. (In Russian.)

1907b. The phenomenon of symbiosis from the point of view of the dynamic equilibrium of cohabiting organisms. *Plant Diseases* 1(1–2): 35–51. (In Russian.)

1910. Preface, "The interrelationships between the gonidia and the fungal component of lichen symbiosis," by A. N. Danilova. *Proceedings of the St. Petersburg Botanical Garden* 10(2): 33–70. (In Russian.)

1912. The lichen *Saccomorpha arenicola mihi*, characterization of the new genus *(Saccomorpha mihi)* and new family *(Saccomorphaceae mihi)*. *Transactions of the Freshwater Station, St. Petersburg Society of Natural Sciences* 3:174–205. (In Russian.)

1921a. The scientific and civic man A. S. Famintsyn. *Proceedings of the Main Botanical Garden of the Russian Soviet Federated Socialist Republic* 20(2): 67–74. (In Russian.)

1921b. The law of dynamic equilibrium in cohabiting and communal plants. *Proceedings of the Main Botanical Garden of the Russian Soviet Federated Socialist Republic* 20(2): 75–121. (In Russian.)

1921c. Lichens as objects of pedagogy and scientific research. *Field Science* 2–3:114–78. (In Russian.)

1922a. The connection between the blue-green alga *Nostoc zetterstedii* Aresch. and the deep-water lichen *Collema ramenskii mihi* nov. sp. *Botanical Materials of the Institute of Spore-bearing Plants of the Main Botanical Garden of the Russian Soviet Federated Socialist Republic* 1(3): 35–46. (In Russian.)

1922b. The new lichen *Pseudoperitheca murmanica mihi* (nov. gen. et sp.) and the peculiar feeding organs in the lichen *Saccomorpha* Elenk. and *Pseudoperitheca* Elenk. *Botanical Materials of the Institute of Spore-bearing Plants of the Main Botanical Garden of the Russian Soviet Federated Socialist Republic* 1(4): 49–56. (In Russian.)

1922c. A new species of the crustose lichen *Leptogium issatschenkii mihi* in the Main Botanical Garden and new sections of the genus *Pseudomallotium mihi*. *Botanical Materials of the Institute of Spore-bearing Plants of the Main Botanical Garden of the Russian Soviet Federated Socialist Republic* 1(5): 65–69. (In Russian.)

1922d. New works in foreign and Russian journals relating to my theory of endoparasitic saprophytism and the law of dynamic equilibrium in the components of the lichenous symbiosis. *Proceedings of the Main Botanical Garden of the Russian Soviet Federated Socialist Republic* 21(1): 65–69. (In Russian.)

1926. The evolution of lower algae and the theory of equivalentogenesis. *Botanical Materials of the Institute of Spore-bearing Plants of the Main Botanical Garden of the Russian Soviet Federated Socialist Republic* 4(1–2): 1–24. (In Russian.)

1936. *Blue-Green Algae of the USSR*. General part. Moscow-Leningrad. (In Russian.)

1939. Invalid "law" of dynamic equilibrium and the theory of equivalentogenesis. *Contemporary Botany* 6–7:113–28. (In Russian.)

1941. The "lichen" and "lichen symbiosis" concepts in light of the studies of Charles Darwin and dialectical materialism. *Reviews of Scientific Research Work of the Biological Science Department of the Soviet Academy of Sciences in 1940*. Moscow, 19–20. (In Russian.)

1975. The "lichen" and "lichen symbiosis" concepts. In *New Systematics of Lower Plants* (Leningrad) 12:3–81. (In Russian.)

PUBLICATIONS BY A. S. FAMINTSYN

1870. The work of Schwendener on lichens. *Transactions of the St. Petersburg Society of Naturalists* 1:39–40. (In Russian.)

1874. *Darwin and His Significance in Biology: Lectures, Delivered in Person in St. Petersburg on February 8, 1874*. St. Petersburg.

1889a. Beitrag zur Symbiose von Algen und Thieren. *Mémoires de l'academie imperial des Sciences de St. Pétersbourg, Série 7*, 36(16): 1–36.

1889b. Danilevsky and Darwinism: Refutation of Darwinism by Danilevsky? *Herald of Europe* 1:616–43. (In Russian.)

1890. The mental life of the simplist life forms as representatives of living matter. In *Transactions of the Eighth Congress of Russian Naturalists and Physicians, General Division*. St. Petersburg, 32–39. (In Russian.)

1891. The symbiosis of algae with animals. *Transactions of the Botanical Laboratory of the Academy of Sciences* 1:1–22. (In Russian.)

1893. The fate of chlorophyll in seeds and sprouts. *Transactions of the Botanical Laboratory of the Academy of Sciences* 5:1–16. (In Russian.)

References

1894. The next problem in biology. *Herald of Europe* 132–53. (In Russian.)

1898. Contemporary natural science and psychology. *God's World* 1–7; 3, 167–99. (In Russian.)

1899. Contemporary natural science and its next problem. *God's World* 12:1–12. (In Russian.)

1907a. The role of symbiosis in the evolution of organisms. *Transactions of the Academy of Sciences, series 8, Physical-Mathematical Division* 20(3): 1–14. (In Russian.)

1907b. The role of symbiosis in the evolution of organisms. *Transactions of the St. Petersburg Society of Natural Science,* vol. 38, issue 1, minutes of session, no. 4, 141–43. (In Russian.)

1912a. The role of symbiosis in the evolution of organisms. *Proceedings of the Academy of Sciences,* ser. 6, 6(11): 51–68. (In Russian.)

1912b. The role of symbiosis in the evolution of organisms. *Proceedings of the Academy of Sciences,* ser. 6, 6(1): 707–14. (In Russian.)

1914. Zoospores in lichens. *Proceedings of the Academy of Sciences,* ser. 6, 8(6): 429–33. (In Russian.)

1916. The role of symbiosis in the evolution of organisms. *Proceedings of the Petrograd Biological Laboratory* 15(3–4): 3–4. (In Russian.)

1918. What are lichens? *Nature* (April–May): 266–82. (In Russian.)

Famintsyn, A. S., and O. Baranetsky. 1867. Zur Entwickelungsgeschichte der Gonidien und Zoosporen: Bildung der Flechten. *Mémoires de l'academie imperial des Sciences de St. Pétersbourg, Série 7,* 11(9): 1–6.

Famintsyn, A. S., and V. Serk. 1915. More on zoospores of lichens: The formation of zoospores in gonidia, related with hyphae. *Proceedings of the Academy of Sciences,* ser. 6, 9(11): 1203–08. (In Russian.)

Filipchenko, A. A. 1937. Ecological concepts of parasitism and parasitism as an independent scientific discipline. In *Problems of General Parasitology.* Leningrad-Moscow, 4–14. (In Russian.)

Filippovich, I. I., E. N. Svetaylo, and K. A. Aliev. 1970. Features and peculiarities found in components of the protein synthesis systems of chloroplasts. In *Functional Biochemistry of Cellular Structure,* 132–42. (In Russian.)

Fox, S., and K. Doze. 1975. *Molecular Evolution and the Origin of Life.* Moscow. (In Russian.)

Gauze, G. G. 1977. *Mitochondrial DNA.* Moscow. (In Russian.)

PUBLICATIONS BY A. G. GENKEL'

1904. Cohabitation in the Plant Kingdom. *Bulletin and Library of Self-Education* 34: 1275–80.

1921. On helotism in lichens. In *Journal of the First All-Russian Congress of Russian Botanists in Petrograd for 1921.* Petrograd. (In Russian.)

1923. On helotism in lichens. *Proceedings of the Biological Scientific Research Institute and Biological Station of Perm University* 1(3–4): 60–64. (In Russian.)

1924. Symbiosis and symbiogenesis. *Man and Nature* 7–8:558–64. (In Russian.)

References

1938. On lichen symbiosis. *Bulletin of the Moscow Society of Natural Research, Biological Division* 47(1): 13–19. (In Russian.)

1946. New observations upon the trifold nature of lichen symbiosis. *Bulletin of the Moscow Society of Natural Research, Biological Division* 51(6): 51–58. (In Russian.)

1974. *Microbiology and the Essentials of Virology.* Moscow. (In Russian.)

1976. [Preface]. In *The Comparative Organellography of Cytoplasm,* A. Frey-Wyssling. Moscow. (In Russian.)

1977. Symbiosis in the plant world. *Advances in Contemporary Biology* 84(4): 138–51. Pt. 1 of a series.(In Russian.)

Genkel', P. A. (reviewer), L. P. Breslavets, B. L. Isachenko, N. A. Komarnitsky, S. Yu. Lipshits, and N. A. Maksipov. 1949. Essays on the history of Russian botanists. *Transactions of the Institute of the History of Natural Science and Technics* 3:419–25. (In Russian.)

Genkel', P. A., and T. T. Plotnikova. 1973. Nitrogen-fixing bacteria in lichens. *Proceedings of the USSR Academy of Sciences, Biological Series* 6:807–13. (In Russian.)

Genkel', P. A., and L. A. Yuzhakova. 1936. Nitrogen-fixing bacteria in lichens. *Proceedings of the Biological Scientific Research Institute and Biological Station of Perm University* 10(9–10): 315–28. (In Russian.)

Gibor, A., and S. Granick. 1964. Plastids and mitochondria: Heritable systems. *Science* 145:890–97.

Gokhlerner, G. B. 1977. The development of aerobic life and the problem of cellular evolution. *Nature* 6:47–57. (In Russian.)

Gollerbach, M. M. 1928. Several additions to the anatomy of the subaqueous lichen *Collema ramenskii* Elenk. *Proceedings of the Main Botanical Garden of the Russian Soviet Federated Socialist Republic* 27(3): 306–13. (In Russian.)

———. 1930. The morphology and biology of *Leptogium issatschenkoi* Elenk. in natural habitat conditions. *Proceedings of the Main Botanical Garden of the USSR* 29(3–4): 300–24. (In Russian.)

Gollerbach, M. M., and A. A. Elenkin. 1938. *Lichens—Their Structure, Life, and Significance.* Leningrad, 3–70. (In Russian.)

Gollerbach, M. M., and V. I. Polyansky. 1951. *Determination of Freshwater Algae,* issue 1, general part. Moscow. (In Russian.)

Gollerbach, M. M., and T. V. Sedova. 1974. Symbiosis by algae. *Botanical Journal* 59(9): 1359–74. (In Russian.)

Gorbunova, N. P. 1956. The interrelationship of fungi and plants in mycorrhizae. *Advances in Contemporary Biology* 62(5): 160–75. Pt. 2 of a series. (In Russian.)

Gordyagin, A. Ya. 1933. Through the history of the botanical cabinet. *Student Reports of Kazan University* 6(1): 46–65. (In Russian.)

Granick, S., and A. Gibor. 1967. The DNA of chloroplasts, mitochondria and centrioles. *Advances / Progress in Nucleic Acid Research and Molecular Biology* 1 / 6:143–86.

Green, P. B. 1964. Cinematic observations on the growth and division of chloroplasts in *Nitella. American Journal of Botany* 51(3): 334–42.

Gromov, B. V. 1976. *The Ultrastructure of Blue-Green Algae.* Leningrad. (In Russian.)

References

Hall, W., and G. Claus. 1963. Ultrastructural studies on the blue-green algal symbiont in *Cyanophora paradoxa* Korschikoff. *Journal of Cell Biology* 19(3): 551–63.

———. 1967. Ultrastructural studies on cyanelles of *Glaucocystis nostochinearum* Itzigsohn. *Journal of Phycology* 3(1): 37–51.

Harley, J. L. 1963. *The Biology of Mycorrhiza.* London.

Hayes, W. 1965. *The Genetics of Bacteria and Their Viruses.* Moscow. (In Russian.)

Henry, S. M. (ed.). 1966–67. *Symbiosis,* vol. 1. Academic Press, New York.

Hewitt, L. 1956. The influence of bacteriophages on the mutability and evolution of bacteria. In *Adaptations of Microorganisms.* Moscow, 422–46. (In Russian).

History and Present State of the Physiology of Plants in Academy Sciences. 1967. [From the laboratory of academician A. S. Famintsyn to the Timiryazev Institute of Physiology]. Moscow. (In Russian.)

Huxley, J. 1963. *Evolution: The Modern Synthesis.* London.

Imshenetsky, A. A. 1945. The nuclear apparatus in bacteria. *Microbiology* 14(2): 65–79. (In Russian.)

———. 1954. The formation of the nucleus in bacterial cells. In *New Data on the Problem of Development in Cellular and Acellular Forms of Living Matter.* Moscow, 197–202. (In Russian.)

Imshenetsky, A. A., G. A. Zavarzin, and V. V. Alferov. 1959. The nucleus in bacteria. *Thesis Proceedings of the Coordinated Conference on the Problems "Difficult Questions of Cytology."* Leningrad, 66–67. (In Russian.)

Iskina, P. E. 1938. The question of nitrogen-fixing bacteria in lichens. *Proceedings of the Biological Science Research Institute of Perm University* 11(5–6): 33–139. (In Russian.)

Ivanov, A. V. 1968. *The Origin of Multicellular Animals.* Leningrad. (In Russian.)

Iwamura, T. 1960. Distribution of nucleic acids among subcellular fractions of *Chlorella. Biochimica et Biophysica Acta* 42(1): 161–63.

Jacob, F., and E. Vollman. 1962. *Sex and Genetics of Bacteria.* Moscow. (In Russian.)

Jinks, J. 1966. *Extrachromosomal Inheritance.* Moscow. (In Russian.)

Jennings, D. H., and D. L. Lee. (eds.). 1975. *Symbiosis.* London.

Kallinikova, V. D. 1974. Evolutionary aspects in the study of kinetoplasts. *Cytology* 16:1191–1202. (In Russian.)

Kamensky, F. M. 1891. *The Appearance of Symbiosis in the Plant Kingdom.* Odessa. (In Russian.)

Kamyshev, N. S. 1957. Kozo-Polyansky. *Botanical Journal* 42(10): 1530–35. (In Russian.)

Katsnel'son, Z. S. 1945. Biotic factors of the environment and their classification. *Journal of General Biology* 6(3): 205–16. (In Russian.)

Keller, B. A. 1933. *Genetics.* Moscow. (In Russian.)

———. 1935. *Botany: Chief Facts and Principles.* Moscow. (In Russian.)

PUBLICATIONS BY L. N. KHAKHINA

1971. The problem of symbiogenesis in the work of Russian scientists. In *Science and Technology,* issue 6. Leningrad, 187–90. (In Russian.)

References

1972a. The significance of integration of organisms at the individual level for the evolutionary process. In *The Organization and Evolution of Life.* Leningrad, 88–92. (In Russian.)

1972b. The impact of the concept of symbiogenesis in the work of B. M. Kozo-Polyansky. In *Science and Technology,* issue 7. Leningrad, 30–33. (In Russian.)

1973a. A history of studies on symbiogenesis. In *Through the History of Biology,* issue 4. Moscow, 63–75. (In Russian.)

1973b. Experimental sources of studies on symbiogenesis (the work of A. S. Famintsyn). In *History and Theory of Evolutionary Studies,* issue 1. Leningrad, 129–41. (In Russian.)

1973c. Theoretical sources of opinions of A. S. Famintsyn on the role of symbiosis in the evolution of organisms. In *History and Theory of Evolutionary Studies,* issue 1. Leningrad, 142–49. (In Russian.)

1973d. Ideas of symbiogenesis and their place in the system of evolutionary statements of A. S. Famintsyn. In *Science and Technology,* issue 8. Leningrad, 181–85. (In Russian.)

1975. The formation of the hypothesis of symbiogenesis by K. S. Merezhkovsky. In *History and Theory of Evolutionary Studies,* issue 3. Leningrad, 5–28. (In Russian.)

Khokhlov, S. S. 1977. *Symbiogenesis: History of the Origin of the Cell and the Organisms by Method of Symbiosis. Two Lectures on Evolutionary Studies.* Saratov. (In Russian.)

Kirk, J. T. 1970. Plastid anatomy. In *Functional Biochemistry of Cellular Structure.* Moscow, 39–51. (In Russian.)

Knorre, A. G. 1937. The extent of parasitism in the animal kingdom. In *Problems of General Parasitology.* Leningrad-Moscow, 21–28. (In Russian.)

Koloss, E. I. 1975. On the development of cells and their methods of division. In *Problems of Evolution* 4:44–69. (In Russian.)

Komarnitsky, N. A. 1945. Lichens. In *A Course in the Lower Plants,* by L. I. Kursanov and N. A. Komarnitsky, chaps. 17 and 18. Moscow-Leningrad, 416–56, 457–64. (In Russian.)

––––––. N. A. 1947. Plant morphology. In *Essays on the History of Russian Botany.* Moscow, 115–75. (In Russian.)

––––––. N. A. 1948. Essay on the history of studies on the lower plants in Russia and the USSR. *Student Reports of Moscow University* 129(6): 71–118. (In Russian.)

Komarnitsky, N. A., and S. Yu. Lipshits. 1945. A. A. Elenkin as botanist. *Bulletin of the Moscow Society of Natural Research, Biological Division* 50(1–2): 123–38. (In Russian.)

Komarov, N. F. 1940. The idea of development and the theory of dynamic equilibrium in Soviet geobotany. *Contemporary Botany* 5–6:288–301. (In Russian.)

Korshikov, A. A. 1924. Protistological notes. *Russian Archive of Protistology* 3:57–74. (In Russian.)

PUBLICATIONS BY B. M. KOZO-POLYANSKY

1921a. The theory of symbiogenesis and "pangenesis, provisional hypothesis." In *Journal of the First All-Russian Congress of Russian Botanists in Petrograd in 1921.* Petrograd. (In Russian.)

References

1921b. Symbiogenesis in the evolution of the plant world. *Bulletin of Experimental Methods* (Voronezh) 4:1–24. (In Russian.)

1923. *The Latest Word in Antidarwinism. Critical Nomogenesis.* Krasnodar. (In Russian.)

1924. *A New Principle of Biology. Essay on the Theory of Symbiogenesis.* Moscow. (In Russian.)

1925a. *Darwinism. The Scheme.* Vologda-Moscow. (In Russian.)

1925b. *Dialectics in Biology.* Rostov-na-Donu–Krasnodar. (In Russian.)

1926. Clarification of several of our positions in Darwinism. In *Dialectics in Nature.* Moscow, 233–64. (In Russian.)

1932. *Introduction to Darwinism.* Voronezh. (In Russian.)

1937. *Fundamental Biogenetic Law from a Botanical Point of View.* Voronezh. (In Russian.)

1948. The modernized system of the plant world. *Transactions of Voronezh University* 15:76–129. (In Russian.)

1965. *A Course in the Higher Plants.* Voronezh. (In Russian.)

Kuprevich, V. F. 1952. The physiological role of mycorrhizae. *Transactions of Complex Scientific Examination of Questions of Practical Forestry* 2(2): 39–46. (In Russian.)

Kursanov, L. I., and N. A. Komarnitsky. 1945. *A Course in Lower Plants.* Moscow. (In Russian.)

Kusakin, O. G., and Ya. I. Starobogatov. 1973. The question of the highest taxonomic categories in the organic world. In *Problems of Evolution*, vol. 3. Moscow, 95–103. (In Russian.)

Kuznetsov, S. I. 1974. Development of the ideas of S. N. Vinogradsky in the field of ecological microbiology. In *The Second Presentation in Honor of S. N. Vinogradsky, 11 October 1972.* Moscow. (In Russian.)

Lee, R. E. 1972. Origin of plastids and the phylogeny of algae. *Nature* 237(5349): 44–46.

Leninger, A. 1966. *Mitochondria.* Moscow. (In Russian.)

———. 1974. *Biochemistry. The Molecular Bases of Structure and Function in Cells.* Moscow. (In Russian.)

Levina, R. E. 1965. Memories of a teacher: On the seventy-fifth birthday of B. M. Kozo-Polyansky. *Botanical Journal* 50(11): 1651–54. (In Russian.)

Limbaugh, C. 1961. Cleaning symbioses. *Scientific American* 205(2): 42–49.

Lipshits, S. Yu. 1950. Russian botanists. A. A. Elenkin. *Biographical-Bibliographical Dictionary*, vol. 3. Moscow, 244–56. (In Russian.)

Lipskaya, A. A., S. B. Ivanova, and M. R. Grabovskaya. 1974. Characteristics of DNA of chloroplasts and blue-green algae. In *Problems of Evolutionary Physiology of Plants.* Leningrad, 68–71. (In Russian.)

Lybishchev, A. A. 1972. Logical systematics. In *Problems of Evolution*, vol. 2. Moscow, 45–68. (In Russian.)

Lyubimenko, V. N. 1916. The transformation of pigments of plastids in living plant tissue. *Transactions of the Academy of Sciences*, ser. 8, Physical-Mathematical Department, 33:1–274. (In Russian.)

———. 1917. The question of physiological independence of plastids. *Journal of the Russian Botanical Society* 2(1–2): 46–56. (In Russian.)

References

————. 1923. *A Course in General Botany*. Berlin. (In Russian.)

————. 1935. Photosynthesis and chemosynthesis in the plant world. Moscow-Leningrad, 5–321. Also: 1963. *Selected Works in Two Volumes,* vol. 1. Kiev, 320–597. (In Russian.)

MacDougall, V. B. 1935. *Ecology of Plants*. Moscow. (In Russian.)

Manoylenko, K. V. 1974. *The Development of Evolutionary Trends in the Physiology of Plants*. Leningrad. (In Russian.)

Manoylenko, K. V., and L. N. Khakhina. 1974. Through the history of development of evolutionary theory in the Academy of Sciences and the contributions of academician A. S. Famintsyn. *Journal of General Biology* 35(2): 308–14. (In Russian.)

PUBLICATIONS BY L. MARGULIS

1968. Evolutionary criteria in Thallophytes: A radical alternative. *Science* 161(3845): 1020–22.

1970. *Origin of Eukaryotic Cells*. Yale University Press, New Haven.

1971a Symbiosis and evolution. *Scientific American* 225(2): 48–57.

1971b. Whittaker's five kingdoms of organisms: Minor revisions suggested by considerations of the origin of mitosis. *Evolution* [USA] 25(1): 242–45.

1974. Five-kingdom classification and the origin and evolution of cells. *Evolutionary Biology* (W. Steere, M. Hecht, and T. Dobzhansky, eds.) 7:45–78. Plenum Press, New York.

1975. Symbiotic theory of the origin of eukaryotic organelles: Criteria for proof. In *Symbiosis*. London.

[1992. *Symbiosis in Cell Evolution: Microbial Communities in the Archean and Proterozoic Eons*. 2d ed. W. H. Freeman, New York.]

Mednikov, B. M. 1975. *Darwinism in the Twentieth Century*. Moscow. (In Russian.)

Meyster, G. K. 1934. *Critical Essay on the Fundamental Concepts of Genetics*. Moscow-Leningrad. (In Russian.)

PUBLICATIONS BY K. S. MEREZHKOVSKY

1883. *Materials Contributing to the Knowledge of Animal Pigments*. St. Petersburg. (In Russian.)

1903. *Morphology of Diatoms*. Kazan. (In Russian.)

1905a. Über Natur und Ursprung der Chromatophoren im Pflanzenreiche. *Biol. Centralbl.* 25(18): 593–604.

1905b. On the occasion of my work on the endochromes of diatoms. In *Transactions of the University of Kazan,* vol. 72, bk. 9, 1–44. (In Russian.)

1906. Principles of endochromes. In *Memoirs of the University of Kazan,* vol. 73, bks. 2, 3, 5–6, pp. 1–176, 177–288, 289–402. (In Russian.)

1909a. *The Theory of Two Plasms as the Basis of Symbiogenesis, A New Study on the Origins of Organisms*. Proceedings of Studies of the Imperial Kazan University. Publishing Office of the Imperial University. (In Russian.)

References

1909b. *A Course on Cryptogamic Plants.* Proceedings of Studies of the Imperial Kazan University. Publishing Office of the Imperial University. (In Russian.)

1910. *A Course on General Botany.* In *Memoirs of the University of Kazan,* vol. 77. (In Russian.)

1920. La Plante considerée comme un complexe symbiotique. *Societé des Sciences Naturelles de l'Ouest de la France, Nantes, Bulletin* 6:17–98.

Mordvilko, A. K. 1936. Ants and aphids. *Nature* 4:44–53. (In Russian.)

Moshkovsky, Sh. D. 1946. Functional parasitology. *Medical Parasitology and Parasitic Disease* 15(4): 26–36. (In Russian.)

———. 1957. The nature of protozoans (Protozoa) and the frontiers of protozoology. *Transactions of the Leningrad Society of Natural Science* 73(4): 129–37. (In Russian.)

Mühlethaler, K., and P. R. Bell. 1962. Untersuchungen über die Kontinuität von Plastiden und Mitochondrien in der Eizelle von *Pteridium aquilinum* (L.) Kühn. *Naturwissenschaften* 49(3): 63–64.

Nägeli, C. 1884. *Mechanisch-physiologische Theorie der Abstammungslehre.* Munich-Leipzig.

Nass, M. M. T. 1969. Uptake of isolated chloroplasts by mammalian cells. *Science* 165:1128–31.

Nass, S. 1969. The significance of the structural and functional similarities of bacteria and mitochondria. *International Review of Cytology* 25:55–117.

Nasyrov, Yu. S. 1975a. Autonomous chloroplasts? *Nature* 11:43–48. (In Russian.)

———. 1975b. *Photosynthesis and the Genetics of Chloroplasts.* Moscow. (In Russian.)

Navashin, S. G. (reviewer) 1892. The symbiosis of algae with animals. In *Review of Botanical Work in Russia in 1891.* St. Petersburg, 81–86. (In Russian.)

———. 1916. Principles of continuity and new methods in the study of the cells of higher plants. *Journal of the Russian Botanical Society* 1(1–2): 1–38. (In Russian.)

———. 1926. *Sex: A Factor in Organic Evolution.* Vologda. (In Russian.)

Nutman, P. S., and Mosse, B. (eds.). 1963. *Symbiotic Associations.* Cambridge.

Odintsova, M. S. 1976. DNA of chloroplasts and mitochondria (structure, replication, physicochemical properties). In *Biological Chemistry,* vol. 10. Moscow. (In Russian.)

Odum, E. 1968. *Ecology.* Moscow. (In Russian.)

Odum, Yu. [Odum, E.]. 1975. *Basic Ecology.* Moscow. (In Russian.)

Oksner, A. N. [1956.] *The Lichen Flora of Ukraine,* vol. 1. Kiev. (In Russian.)

———. 1974. *Identification of Lichens of the USSR,* issue 2, general part. Leningrad. (In Russian.)

Oparin, A. I. 1941. *The Emergence of Life on Earth.* Moscow-Leningrad. (In Russian.)

———. 1976. *The Problem of the Origin of Life.* Moscow. (In Russian.)

Oschman, J. L., and P. Gray. 1965. A study of the fine structure of *Convoluta roscoffensis* and its endosymbiotic algae. *Transactions of the American Microscop. Society* 84:368–75.

Ostroumov, S. A. 1973. What can be said on the similarity of mitochondria, chloroplasts, and prokaryotes? *Nature* 3:21–29. (In Russian.)

References

Pakhomova, M. V. 1972. The DNA of Algae. In *The Structure of DNA and the Systematic Position of Organisms.* Moscow, 176–95. (In Russian.)

――――. 1974. Nucleic acid of blue-green algae. In *Topical Problems of the Biology of Blue-Green Algae.* Moscow, 104–13. (In Russian.)

Parnes, V. A. 1972. *Antone de Bary.* Moscow. (In Russian.)

Pavlovsky, E. N. 1934. Organisms as inhabitants of the environment. *Nature* 1:80–91. (In Russian.)

Perfil'ev, B. V. 1914. The study of symbiosis. *Microbiology* 1(3–5): 209–24. (In Russian.)

Pinevich, V. V., T. P. Soidla, S. B. Ivanova, T. P. Levitina, and A. A. Lipskaya. 1974. The origin of the photosynthetic systems of eukaryotes (The question of the symbiotic hypothesis of the origin of chloroplasts). *Cytology* 16:124–36. (In Russian.)

Pinevich, V. V., and V. E. Vasil'eva. 1972. Carotenoids of blue-green algae: Their significance for understanding evolution and mechanisms of photosynthesis. *Bulletin of Leningrad University* 21:105–22. (In Russian.)

Pinus, E. A. 1973. Autonomous mitochondria. In *Problems of the Origins and Essence of Life.* Moscow, 246–50. (In Russian.)

Polyansky, V. I. 1947. *Selected Chapters on Darwinism.* Leningrad. (In Russian.)

――――. 1957. Introduction to "In Eastern Sayan with V. L. Komarov," by A. A. Elenkin. *Transactions of the Institute of the History of Natural Science and Technics* 16:253–334. (In Russian.)

Polyansky, V. I., and A. A. Strelkov. 1935. The influence of the faunas of infusorian infections on animal growth. *Transactions of the Petergof Biological Institute, Leningrad State University* 13–14:68–87. (In Russian.)

Preer, J. R., L. B. Preer, and A. Jurand. 1974. Kappa and other endosymbionts in *Paramecium aurelia. Bacteriological Review* 38:113–63.

Rabotnov, T. A. 1969. Consortia. *Bulletin of the Moscow Society of Natural Research, Biological Division* 74(4): 109–16. (In Russian.)

――――. 1974. Consortia as the structural units of biogeocenoses. *Nature* 2:26–35. (In Russian.)

Raykov, B. E. 1957. On the history of Darwinism in Russia. *Transactions of the Institute of the History of Natural Science and Technics* 16:3–33. (In Russian.)

Raven, P. H. 1970. A multiple origin for plastids and mitochondria. *Science* 169(3946): 641–46.

Reinke, J. 1873. *Morphologische Abhandlungen.* Leipzig.

――――. 1894 [and 1895]. Abhandlungen über Flechten. *Jahrbuecher fuer Wissenschaftliche Botanik* 1894:495–542; 1895:395–486.

Ris, H., and W. Plaut. 1962. Ultrastructure of DNA-containing areas in the chloroplast of *Chlamydomonas. Journal of Cell Biology* 13(3): 383–92.

Robertis, E., V. Novinsky, and F. Saes. 1973. *Biology of Cells.* Moscow. (In Russian.)

Roodyn, D., and D. Wilkie. 1970. *Biogenesis of Mitochondria.* Moscow. (In Russian.)

Rubin, B. A., and E. V. Artsekhovskaya. 1968. *Biochemical and Physiological Immunity of Plants.* Moscow. (In Russian.)

Rutsky, I. A. 1959. Colorful memories of Professor B. M. Kozo-Polyansky. *Bulletin of the*

References

Society of Natural Science of the University of Voronezh 11:5–11. (In Russian.)

Rutten, M. 1973. *The Origin of Life: By Natural Causes.* Moscow. (In Russian.)

PUBLICATIONS BY V. L. RYZHKOV

1924. *The Rudiments of Biology: History of Species and Individuals.* Kharkov. (In Russian.)

1927. *News in Biology: Popular Essays.* Kharkov. (In Russian.)

1939. The labile state of genes, genomes, and cytoplasm. *Advances in Contemporary Biology* 11(2): 348–63. (In Russian.)

1942. Problems in the evolution of ultraviruses. *Microbiology* 11(4): 149–59. (In Russian.)

1960. Alleged infectious hereditary of bacteria. *Questions of Virology* 5:515–20. (In Russian.)

1961. Short essays on the history of study of viruses. *Transactions of the Institute of the History of Natural Science and Technology* 36:315–25. (In Russian.)

1964. Symbiosis at the molecular level. In *Thesis Lectures on the Nature of Viruses at the Conference Celebrating the Centenary of the Birth of D. I. Ivanovsky.* Moscow, 12–15. (In Russian.)

1965. Symbiosis at the molecular level. *Advances in Contemporary Biology* 59(3): 385–98. (In Russian.)

1966. Intracellular symbiosis. *Nature* 3:9–17. (In Russian.)

1976. What's old in new genetics? *Philosophical Questions* 3:89–95. (In Russian.)

Sagan, L. [L. Margulis]. 1967. On the origin of mitosing cells. *Journal of Theoretical Biology* 14(3): 225–75.

Samtsevich, S. A. 1969. The significance of tubercle-forming bacteria as symbionts of leguminous plants. In *The Role of Microorganisms in the Nourishment of Plants and Soil Fertility.* Minsk, 4–56. (In Russian.)

Sapegin, A. A. 1913. *Research on the Individualness of the Plastid.* Odessa. (In Russian.)

Savich, V. P. 1944. In memory of A. A. Elenkin (1873–1942). *Soviet Botany* 1:60–63. (In Russian.)

Schwendener, S. 1869. *Die Algentypen der Flechtengonidien.* Basel.

Sedova, T. L. 1977. *Basic Cytology of Algae.* Leningrad. (In Russian.)

Sedzher, R. 1975. *Cytological Genes and Organelles.* Moscow. (In Russian.)

Senchenkova, E. M. 1960. Andrei Sergeevich Famintsyn (on the fortieth anniversary of his death). *Botanical Journal* 45(2): 309–18. (In Russian.)

———. 1969. The development of ideas on the origin and evolution of photosynthetic plants. In *Problems of Plant Physiology: Historical Essays.* Moscow, 111–63. (In Russian.)

———. 1973. The development of ideas on autonomous chloroplasts. In *Through the History of Biology,* issue 4. Moscow, 41–62. (In Russian.)

———. 1974. The discovery of chromatography and the Academy of Sciences. *Nature* 5:93–101. (In Russian.)

References

Shaposhnikov, G. Kh. 1975. Living systems with a low degree of integrity. *Journal of General Biology* 36(3): 323–35. (In Russian.)

Shemakhanova, N. M. 1962. *Mycotrophic Tree Species.* Moscow. (In Russian.)

Shmal'gauzen, I. I. 1969. *Problems of Darwinism.* Leningrad. (In Russian.)

Skryabin, K. I. 1923. *Symbiosis and Parasitism in Nature.* Petrograd. (In Russian.)

Smirnov, L. A. 1952. A discovery made twice. *Botanical Journal* 37(5–6): 896–98. (In Russian.)

Stanier, R. J. 1970. Some aspects of the biology of cells and their possible evolutionary significance. In *Prokaryotic and Eukaryotic Cells,* Symposium of the Society of General Microbiology. Cambridge.

Starr, M. P. 1975. A generalized scheme for classifying organismic associations. In *Symbiosis, Symposium of the Society of Experimental Biology* 29:1–20. London.

Strogonov, B. P. 1974. The physiology of plants in the USSR Academy of Sciences (origin and development of plant physiology). *Physiology of Plants* 21(3): 445–54. (In Russian.)

Sukachev, V. N., and N. V. Dylis. 1964. *The Basics of Forest Biogeocenology.* Moscow. (In Russian.)

Sukhov, K. S. 1942. *Plant Viruses and Insect Vectors.* Moscow. (In Russian.)

Svetaylo, E. N. 1973. Several biochemical aspects of the problem of the origin of chloroplasts. In *Problems of the Origin and Essence of Life.* Moscow, 238–45. (In Russian.)

Sytnik, K. M., and A. V. Gordetsky. 1974. Plant physiology in the light of contemporary evolutionary theory. In *Problems of the Evolutionary Physiology of Plants.* Leningrad, 12–14. (In Russian.)

Takhtadzhyan, A. L. 1950. B. M. Kozo-Polyansky. On the sixtieth anniversary of his birth. *Botanical Journal* 35(4): 416–31. (In Russian.)

———. 1973. Four kingdoms of the organic world. *Nature* 2:22–32. (In Russian.)

———. 1974. Plants in the system of organisms. In *The Life of Plants,* vol. 1. Moscow, 49–57. (In Russian.)

———. 1976. The system of the organic world. *Great Soviet Encyclopedia,* 3d ed., vol. 23, 466–68. (In Russian.)

Taylor, D. L. 1970. Chloroplasts as symbiotic organelles. *International Review of Cytology* 27:29–64.

Taylor, F. J. R. 1976. Autogenous theories for the origin of eukaryotes. *Taxon* 25(4): 337–90.

Tchaikovsky, Yu. V. 1977. Genetic integration of cellular structure as a factor in evolution. *Journal of General Biology* 38(6): 823–35. (In Russian.)

Terekhin, E. S. 1965. Lichens: Their origin and the role of symbiogenesis in evolution. *Science and Life* 1:94–96. (In Russian.)

Timiryazev, K. A. 1887. Darwinism refuted? *Russian Thought* 5(2): 145–80 and 6(2): 1–14. (Also in *Works,* vol. 7. Moscow, 1939, 263–326.) (In Russian.)

———. 1888. Plants: A sphinx. In *Public Lectures and Speeches,* 209–29. (Also in *Works,* vol. 1. Moscow, 1937, 294–313.) (In Russian.)

———. 1889. A strange model of scientific criticism. *Russian Thought* 3(2): 90–102. (Also in *Works,* vol. 7. Moscow, 1939, 425–43. (In Russian.)

References

————. 1922. Symbiosis. *Encyclopaedic Dictionary "The Pomegranate,"* vol. 38, 591–96. (In Russian.)

Trass, Kh. Kh. 1973. Lichens. *Nature* 3:10–20. (In Russian.)

————. 1976. *Geobotany: History and Contemporary Tendencies of Development.* Leningrad. (In Russian.)

Trench, R. K., R. W. Green, and B. G. Bystrom. 1969. Chloroplasts as functional organelles in animal tissues. *Journal of Cell Biology* 42(2): 404–17.

Tswett, M. [1896.] Etudes des physiologie cellulaire Pt. 2, Les chloroplasts. *Archives des Sciences* (Geneva) 2(7): 228–60, 339–48, 467–86, 565–74. (In Russian.)

Urbanke, A. 1973. *Revolucja naukowa w biologii.* Warsaw.

Varming, E. 1901. *Ecological Geography of Plants.* Moscow.

Veisman, A. 1918. *Lectures on Evolutionary Theory.* Petrograd. (In Russian.)

Wallin, I. E. 1927. *Symbionticism and the Origin of Species.* Williams and Wilkins, Baltimore.

Werz, G., and G. Kellner. 1968. Isolierung und electronenmikroskopische Charakterisierung von Desoxyribonucleinsäure aus Chloroplasten kernloser *Acetabularia*—Zellen. *Zeitschrift fuer Naturforschung* 23b:1018–19.

Whittaker, R. H. 1969. New concepts of kingdoms of organisms. *Science* 163:150–60.

Yablokov, A. V., and A. G. Yusufov. 1976. *Evolutionary Studies.* Moscow. (In Russian.)

Yaroshenko, P. L. 1975. *General Biogeography.* Moscow. (In Russian.)

Zablen, L. B., M. S. Kissil, C. R. Woese, and D. E. Bueton. 1975. Phylogenetic origin of the chloroplast and prokaryotic nature of its ribosomal RNA. *Proceedings of the National Academy of Sciences* [USA] 72(6): 2418–22.

Zagoskin, N. P. (ed.). 1904. *Biographical Dictionary of Teachers and Professors of Kazan University.* Publishing House of the Imperial University, Kazan, 435–40. (In Russian.)

Zavadsky, K. M. 1958. Understanding progress in organic nature. In *Problems of Progress in Nature and Society.* Leningrad, 79–120. (In Russian.)

————. 1967. Problems of progress in living nature. *Philosophical Questions* 9:124–36. (In Russian.)

————. 1968. *Species and Speciation.* Leningrad. (In Russian.)

————. 1971. Research on the driving forces of arogenesis. *Journal of General Biology* 5:515–29. (In Russian.)

————. 1973. *The Development of Evolutionary Theory after Darwin (1859 to the 1920s).* Leningrad. (In Russian.)

Zavadsky, K. M., and E. I. Kolchinsky. 1977. *The Evolution of Evolution.* Leningrad. (In Russian.)

Zhukova, G. Ya. 1975. The problem of the origin and evolution of plastids in the light of facts of plant embryology. *Botanical Journal* 60(5): 713–38. (In Russian.)

Zil'ber, L. A. 1952. Symbiosis of viruses and microbes. *Advances in Contemporary Biology* 33(1): 81–100. (In Russian.)

Zotin, A. I., N. D. Ozernyuk, A. A. Zotin, and V. A. Konoplev. 1975. A possible way in which prokaryotes originated. *Journal of General Biology* 2:163–72. (In Russian.)

ЛИТЕРАТУРА

А л е к с а н д р о в В. Г. К вопросу о возникновении зеленых пластид в растительных клетках. — Бот. журн., 1950, т. 35, № 5, с. 475—481.

А н т о н о в А. С. Строение ДНК и положение организмов в системе. — Усп. соврем. биол., 1969, т. 68, вып. 3, с. 299—317.

А н т о н о в А. С. Геносистематика. Достижения, проблемы и перспективы. — Усп. соврем. биол., 1974, т. 77, вып. 2, с. 31—47.

А ф а н а с ь е в В. А. Паразитизм и симбиоз. — В кн.: Проблемы общей паразитологии. Л.—М., 1937, с. 15—20.

Б а з и л е в с к а я Н. А. Творческая деятельность Б. М. Козо-Полянского. — Тр. Ин-та истории естествознания и техники, 1959, т. 23, № 4. с. 324—357.

Б а з и л е в с к а я Н. А., Б е л о к о н ь И. П., Щ е р б а к о в а А. А. Краткая история ботаники. М., 1968. 309 с.

Б а р а н е ц к и й О. О самостоятельной жизни гонидиев лишаев. — Тр. I съезда русских естествоиспыт. и врачей. Отдел бот. СПб., 1868, с. 45—59.

Б е к л е м и ш е в В. Н. Основы сравнительной анатомии беспозвоночных. Т. 1. М., 1964. 432 с.

Б е к л е м и ш е в В. Н. Организм и сообщество. — В кн.: Биоценологические основы сравнительной паразитологии. М., 1970а, с. 26—42.

Б е к л е м и ш е в В. Н. Паразитизм членистоногих на наземных позвоночных. — В кн.: Биоценологические основы сравнительной паразитологии. М., ·1970б, с. 250—260.

Б е л о з е р с к и й А. Н. Нуклеиновые кислоты и их связь с эволюцией, филогенией и систематикой организмов. Ташкент, 1969. 38 с.

Б е л о з е р с к и й А. Н., А н т о н о в А. С., М е д н и к о в Б. М. — Вступительная статья в кн.: Строение ДНК и положение организмов в системе. М., 1972, с. 3—16.

Б е л о з е р с к и й А. Н., М е д н и к о в Б. М. Нуклеиновые кислоты и систематика организмов. М., 1972. 48 с.

Б е р г Л. С. Теории эволюции. Пг., 1922. 119 с.

Б е р и д з е Т. Г., О д и н ц о в а М. С. Дезоксирибонуклеиновая кислота цитоплазматических структур: пластид и митохондрий. — В кн.: Успехи биологической химии. Т. 10. М., 1969, с. 36—63.

Б и о г р а ф и ч е с к и й с л о в а р ь профессоров и преподавателей СПб. университета. Т. II. СПб., 1898, с. 33—36.

Б и о г р а ф и ч е с к и й с л о в а р ь преподавателей и профессоров Казанского университета. Под ред. Н. П. Загоскина. Т. I. Казань, 1904, с. 435—440.

Б л а г о в е щ е н с к и й А. В. Б. М. Козо-Полянский. — Бюл. Глав. бот. сада СССР, 1957, вып. 28, с. 123—124.

Б л я х е р Л. Я. Проблема наследования приобретенных признаков. М., 1971. 274 с.

Б о р о д и н И. П. Андрей Сергеевич Фаминцын (1835—1918). Некролог. — Журн. Рус. бот. о-ва, 1919, т. 4, № 1, с. 132—151.

Б р е с л а в е ц Л. П. История вопроса о происхождении хлоропластов. — Тр. Ин-та истории естествознания и техники, 1959, т. 23. История биол. наук, вып. 4, с. 257—288.

Б р е с л а в е ц Л. П. Современное представление о происхождении пластид. — Изв. АН СССР. Сер. биол., 1963, т. 1, с. 91—98.

Б у т е н к о Р. Г. От свободноживущей клетки — к растению. М., 1971. 95 с.

В а р м и н г Е. Ойкологическая география растений. М., 1901, 538 с.

В е й с м а н А. Лекции по эволюционной теории. Пг., 1918. 359 с.

Г а у з е Г. Г. Митохондриальная ДНК. М., 1977. 287 с.

Г е н к е л ь А. Г. Сожительство в царстве растений. — Вестник и библиотека самообразования, 1904, № 34, с. 1275—1280.

Г е н к е л ь А. Г. К илотизму (=хелотизму) у лишайников. — В кн.: Дневник I Всероссийского съезда русских ботаников в Петрограде в 1921 году. Пг., 1921, с. 81.

Г е н к е л ь А. Г. О хелотизме у лишайников. — Изв. Биол. НИИ и биол. станции при Перм. ун-те. Пермь, 1923, т. 1, вып. 3—4, с. 60—64.

Г е н к е л ь А. Г. Симбиоз и симбиогенез. — Человек и природа, 1924, № 7—8, с. 558—564.

Г е н к е л ь П. А. О лишайниковом симбиозе. — Бюл. МОИП. Отд. биол., 1938, т. 47, вып. 1, с. 13—19.

Г е н к е л ь П. А. Новые наблюдения по тройной природе лишайникового симбиоза. — Бюл. МОИП. Отд. биол., 1946, т. 51, вып. 6, с. 51—58.

Г е н к е л ь П. А. [Рец.] Л. П. Бреславец, Б. Л. Исаченко, Н. А. Комарницкий, С. Ю. Липшиц, Н. А. Максимов. Очерки по истории русской ботаники. — Тр. Ин-та истории естествознания и техники, 1949, т. 3, с. 419—425.

Г е н к е л ь П. А. Микробиология с основами вирусологии. М., 1974. 271 с.

Г е н к е л ь П. А. [Предисловие]. — Фрей-Висслинг А. Сравнительная органеллография цитоплазмы. М., 1976. 144 с.

Г е н к е л ь П. А. О симбиозе в растительном мире. — Усп. соврем. биол., 1977, т. 84, вып. 1 (4), с. 138—151.

Г е н к е л ь П. А., П л о т н и к о в а Т. Т. Азотфиксирующие бактерии в лишайниках. — Изв. АН СССР. Сер. биол., 1973, т. 6, с. 807—813.

Г е н к е л ь П. А., Ю ж а к о в а Л. А. Азотфиксирующие бактерии в лишайниках. — Изв. Биол. НИИ и биол. станции при Перм. ун-те, (Пермь), 1936, т. 10, вып. 9—10, с. 315—328.

Г о л л е р б а х М. М. Некоторые дополнения к анатомии подводного лишайника Collema ramenskii Elenk. — Изв. Глав. бот. сада РСФСР, 1928, т. XXVII, вып. 3, с. 306—313.

Г о л л е р б а х М. М. К морфологии и биологии Leptogium issatschenkoi Elenk. в естественных условиях обитания. — Изв. Глав. бот. сада СССР, 1930, т. XXIX, вып. 3—4, с. 300—324.

Г о л л е р б а х М. М., Е л е н к и н А. А. Лишайники, их строение, жизнь и значение. Л., 1938, с. 3—70.

Г о л л е р б а х М. М., П о л я н с к и й В. И. Определитель пресноводных водорослей. Вып. 1. Общая часть. М., 1951. 200 с.

Г о л л е р б а х М. М., С е д о в а Т. В. Симбиоз у водорослей. — Бот. журн., 1974, т. 59, № 9, с. 1359—1374.

Г о р б у н о в а Н. П. О взаимоотношениях гриба и растений в микоризах. — Усп. соврем. биол., 1956, т. 62, вып. 2 (5), с. 160—175.

Г о р д я г и н А. Я. Из истории ботанического кабинета. — Учен. зап. Казан. ун-та, 1933, т. 6, вып. 1, с. 46—65.

Г о х л е р н е р Г. Б. Развитие аэробной жизни и проблемы клеточной эволюции. — Природа, 1977, № 6, с. 47—57.

Г р о м о в Б. В. Ультраструктура синезеленых водорослей. Л., 1976. 93 с.

Д а д и н г т о н К. Эволюционная ботаника. М., 1972. 307 с.

Д а ж о Р. Основы экологии. М., 1975. 415 с.

Данилов А. Н. О взаимоотношениях между гонидиями и грибным компонентом лишайникового симбиоза. — Изв. СПб. бот. сада, 1910, т. X, вып. 2, с. 33—70.

Данилов А. Н. Симбиоз как фактор эволюции. — Изв. Глав. бот. сада РСФСР, 1921, т. XX, вып. 2, с. 122—136.

Данилов А. Н. Nostoc в симбиозе. — Рус. архив протистологии, 1927, т. VI, вып. 1—4, с. 83—92.

Данилов А. Н. Введение в синтез лишайника Leptogium issatschenkoi Elenk. — Изв. Глав. бот. сада СССР, 1929, т. XXVIII, вып. 3—4, с. 225—264.

Данилов А. Н. Лишайниковый симбиоз. — Природа, 1933, № 11, с. 34—44.

Дарвин Ч. Происхождение видов. — Соч. Т. 3. М.—Л., 1939, с. 253—680.

Джинкс Дж. Нехромосомная наследственность. М., 1966. 288 с.

Догель В. А. Общая протистология. М., 1951. 603 с.

Догель В. А. Общая паразитология. Л., 1962. 464 с.

Дубинин Н. П. Общая генетика. М., 1976. 590 с.

Еленкин А. А. Факультативные лишайники. — Протоколы заседаний. Тр. СПб. о-ва естествоиспыт., 1901, т. XXXII, вып. 1, № 6, с. 261—269.

Еленкин А. А. К вопросу о «внутреннем сапрофитизме» («эндосапрофитизме») у лишайников. — Изв. СПб. бот. сада, 1902, т. II, вып. 3, с. 65—84.

Еленкин А. А. Новые наблюдения над явлениями эндосапрофитизма у лишайников. — Изв. СПб. бот. сада, 1904, т. IV, вып. 2, с. 25—39.

Еленкин А. А. Отношение лишайникового симбиоза к эволюции организмов. — Тр. СПб. о-ва естествоиспыт., 1907а, т. XXXVIII, вып. 1, № 4, с. 160—175.

Еленкин А. А. Явления симбиоза с точки зрения подвижного равновесия сожительствующих организмов. — Болезни растений, 1907б, т. 1, № 1—2, с. 35—51.

Еленкин А. А. [Предисловие к статье А. Н. Данилова «О взаимоотношениях между гонидиями и грибным компонентом лишайникового симбиоза»]. — Изв. СПб. бот. сада, 1910, т. X, вып. 2, с. 33—70.

Еленкин А. А. О лишайнике Saccomorpha arenicola mihi, образующем новый род (Saccomorpha mihi) и новое семейство (Saccomorphaceae mihi). — Тр. Пресноводной станции Пб. о-ва естествоиспыт., 1912, т. 3, с. 174—205.

Еленкин А. А. Научная и общественная деятельность А. С. Фаминцына. — Изв. Глав. бот. сада РСФСР, 1921а, т. XX, вып. 2, с. 67—74.

Еленкин А. А. Закон подвижного равновесия в сожительствах и сообществах растений. — Изв. Глав. бот. сада РСФСР, 1921б, т. XX, вып. 2, с. 75—121.

Еленкин А. А. Лишайники как объект педагогики и научного исследования. — Экскурсионное дело, 1921в, № 2—3, с. 114—178.

Еленкин А. А. О связи между синезеленой водорослью Nostoc zetterstedtii Aresch. и глубоководным лишайником Collema (?) ramenskii mihi nov. sp. — Бот. материалы Ин-та споровых растений Глав. бот. сада РСФСР, 1922а, т. I, вып. 3, с. 35—46.

Еленкин А. А. О новом лишайнике Pseudoperitheca murmanica mihi (nov. gen. et sp.) и о своеобразных органах питания у лишайников Saccomorpha Elenk. и Pseudoperitheca Elenk. — Бот. материалы Ин-та споровых растений Глав. бот. сада РСФСР, 1922б, т. I, вып. 4, с. 49—56.

Еленкин А. А. О новом виде слизистого лишайника Leptogium issatschenkii mihi в Глав. ботаническом саду и новой секции этого рода Pseudomallotium mihi. — Бот. материалы Ин-та споровых растений Глав. бот. сада РСФСР, 1922в, т. I, вып. 5, с. 65—69.

Еленкин А. А. Новейшие работы в иностранной и русской литературе, имеющие отношение к моей теории эндопаразитосапрофитизма и закону подвижного равновесия компонентов лишайникового симбиоза. — Изв. Глав. бот. сада РСФСР, 1922г, т. XXI, вып. 1, с. 65—69.

Еленкин А. А. Эволюция низших водорослей и теория эквивалентогенеза. — Бот. материалы Ин-та споровых растений Глав. бот. сада
РСФСР, 1926, т. 4, вып. 1—2, с. 1—24.

Еленкин А. А. Синезеленые водоросли СССР. Общая часть. М.—Л., 1936.
684 с.

Еленкин А. А. Несостоятельность «закона» подвижного равновесия и
теории эквивалентогенеза. — Сов. ботаника, 1939, № 6—7, с. 113—128.

Еленкин А. А. Понятия «лишайник» и «лишайниковый симбиоз»
в свете учения Ч. Дарвина и диалектического материализма. —
В кн.: Рефераты научно-исследовательских работ Отделения биоло-
гических наук АН СССР за 1940 г. М., 1941, с. 19—20.

Еленкин А. А. Понятия «лишайник» и «лишайниковый симбиоз». —
В кн.: Новости систематики низших растений. Т. 12. Л., 1975,
с. 3—81.

Жакоб Ф., Вольман Э. Пол и генетика бактерий. М., 1962. 475 с.

Жукова Г. Я. Проблема происхождения и эволюции пластид в свете
данных эмбриологии растений. — Бот. журн., 1975, т. 60, № 5,
с. 713—738.

Завадский К. М. К пониманию прогресса в органической природе. —
В кн.: Проблемы развития в природе и обществе. Л., 1958, с. 79—120.

Завадский К. М. Проблема прогресса живой природы. — Вопр. филос.,
1967, № 9, с. 124—136.

Завадский К. М. Вид и видообразование. Л., 1968, 403 с.

Завадский К. М. К исследованию движущих сил арогенеза. — Журн.
общ. биол., 1971, № 5, с. 515—529.

Завадский К. М. Развитие эволюционной теории после Дарвина (1859—
1920-е годы). Л., 1973. 423 с.

Завадский К. М., Колчинский Э. И. Эволюция эволюции. Л., 1977.
235 с.

Зильбер Л. А. О симбиозе вирусов и микробов. — Усп. соврем. биол.,
1952, т. 33, вып. 1, с. 81—100.

Зотин А. И., Озернюк Н. Д., Зотин А. А., Коноплев В. А. Воз-
можный путь происхождения прокариот. — Журн. общ. биол., 1975,
№ 2, с. 163—172.

Иванов А. В. Происхождение многоклеточных животных. Л., 1968. 287 с.

Имшенецкий А. А. О ядерном аппарате у бактерий. — Микробиология,
1945, т. 14, вып. 2, с. 65—79.

Имшенецкий А. А. О формировании ядер в бактериальной клетке. —
В кн.: Новые данные по проблеме развития клеточных и неклеточ-
ных форм живого вещества. М., 1954, с. 197—202.

Имшенецкий А. А., Заварзин Г. А., Алферов В. В. О ядре
у бактерий. — Тез. докл. на координационном совещании по проблеме «Узловые вопросы цитологии». Л., 1959, с. 66—67.

Искина Р. Е. К вопросу об азотфиксирующих бактериях в лишайниках. — Изв. Биол. НИИ при Перм. ун-те. Пермь, 1938, т. 11 вып. 5—
6, с. 33—139.

История и современное состояние физиологии растений в Академии
наук (от Лаборатории акад. А. С. Фаминцына до Института физиологии им. К. А. Тимирязева). М., 1967. 371 с.

Каллиникова В. Д. Эволюционные аспекты в изучении кинетопласта. —
Цитология, 1974, т. 16, с. 1191—1202.

Каменский Ф. М. О явлениях симбиоза в растительном царстве.
Одесса, 1891. 17 с.

Камышев Н. С. Б. М. Козо-Полянский. — Бот. журн., 1957, т. 42, № 10,
с. 1530—1535.

Кацнельсон З. С. Биотические факторы среды и их классификация. —
Журн. общ. биол., 1945, т. 6, № 3, с. 205—216.

Келлер Б. А. Генетика. М., 1933. 120 с.

Келлер Б. А. Ботаника. Главные факты и закономерности. М., 1935.
472 с.

К и р к Дж. Т. О. Автономия пластид. — В кн.: Функциональная биохимия клеточных структур. М., 1970, с. 39—51.

К н о р р е А. Г. Распространение паразитизма в животном царстве. — В кн.: Проблемы общей паразитологии. Л.—М., 1937, с. 21—28.

К о з о - П о л я н с к и й Б. М. Теория симбиогенезиса и «пангенезис, временная гипотеза». — В кн.: Дневник I Всероссийского съезда русских ботаников в Петрограде в 1921 году. Пг., 1921а, с. 101.

К о з о - П о л я н с к и й Б. М. Симбиогенезис в эволюции растительного мира. — Вестн. опыт. дела. Воронеж, 1921б, № 4, с. 1—24.

К о з о - П о л я н с к и й Б. М. Последнее слово антидарвинизма. Критика номогенеза. Краснодар, 1923. 131 с.

К о з о - П о л я н с к и й Б. М. Новый принцип биологии. Очерк теории симбиогенеза. М., 1924. 147 с.

К о з о - П о л я н с к и й Б. М. Дарвинизм. Схема. Вологда—Москва, 1925а. 133 с.

К о з о - П о л я н с к и й Б. М. Диалектика в биологии. Ростов н/Д—Краснодар, 1925б. 93 с.

К о з о - П о л я н с к и й Б. М. К выяснению некоторых наших позиций в дарвинизме. — В кн.: Диалектика в природе. М., 1926, с. 233—264.

К о з о - П о л я н с к и й Б. М. Введение в дарвинизм. Воронеж, 1932. 151 с.

К о з о - П о л я н с к и й Б. М. Основной биогенетический закон с ботанической точки зрения. Воронеж, 1937. 254 с.

К о з о - П о л я н с к и й Б. М. Модернизации системы растительного мира. — Тр. Воронеж. ун-та, 1948, № 15, с. 76—129.

К о з о - П о л я н с к и й Б. М. Курс систематики высших растений. Воронеж, 1965. 407 с.

К о л о с с Е. И. О развитии клетки и способах ее деления. — В кн.: Проблемы эволюции. Т. 4. М., 1975, с. 44—69.

К о м а р н и ц к и й Н. А. Лишайники. — В кн.: К у р с а н о в Л. И., К о м а р н и ц к и й Н. А. Курс низших растений. Гл. XVII и XVIII. М.—Л., 1945, с. 416—456, 457—464.

К о м а р н и ц к и й Н. А. Морфология растений. — В кн.: Очерки по истории русской ботаники. М., 1947, с. 115—175.

К о м а р н и ц к и й Н. А. Очерк истории изучения низших растений в России и СССР. — Учен. зап. Моск. ун-та, 1948, т. 129, вып. 6, с. 71—118.

К о м а р н и ц к и й Н. А., Л и п ш и ц С. Ю. А. А. Еленкин как ботаник. — Бюл. МОИП. Отд. биол., 1945, т. 50, вып. 1, с. 123—138.

К о м а р о в Н. Ф. Идея развития и теория подвижного равновесия в советской геоботанике. — Сов. ботаника, 1940, № 5—6, с. 288—301.

К о р ш и к о в А. А. Протистологические заметки. — Рус. архив протистологии, 1924, т. 3, с. 57—74.

К у з н е ц о в С. И. Развитие идей С. Н. Виноградского в области экологической микробиологии. — В кн.: Второе чтение им. С. Н. Виноградского 11.10.1972. М., 1974. 62 с.

К у п р е в и ч В. Ф. О физиологической роли микоризы. — Тр. Компл. науч. эксп. по вопр. полезащит. лесоразв., 1952, т. 2, № 2, с. 39—46.

К у р с а н о в Л. И., К о м а р н и ц к и й Н. А. Курс низших растений. М., 1945. 488 с.

К у с а к и н О. Г., С т а р о б о г а т о в Я. И. К вопросу о наивысших таксономических категориях органического мира. — В кн.: Проблемы эволюции. Т. 3. М., 1973, с. 95—103.

Л е в и н а Р. Е. Памяти учителя. К 75-летию со дня рождения Б. М. Козо-Полянского. — Бот. журн., 1965, т. 50, № 11, с. 1651—1654.

Л е н и н д ж е р А. Митохондрия. М., 1966. 316 с.

Л е н и н д ж е р А. Биохимия. Молекулярные основы структуры и функции клетки. М., 1974. 957 с.

Л и п с к а я А. А., И в а н о в а С. Б., Г р а б о в с к а я М. Р. Характеристика ДНК хлоропластов и синезеленых водорослей. — В кн.: Проблемы эволюционной физиологии растений. Л., 1974, с. 68—71.

144

Л и п ш и ц С. Ю. Русские ботаники. А. А. Еленкин. — Биографо-библиографический словарь. Т. III. М., 1950, с. 244—256.

Л ю б и м е н к о В. Н. О превращениях пигментов пластид в живой ткани растения. — Зап. Акад. наук. Сер. VIII, физ.-мат. отд-ние, 1916, т. XXXIII, с. 1—274.

Л ю б и м е н к о В. Н. К вопросу о физиологической самостоятельности пластид. — Журн. Рус. бот. о-ва, 1917, т. 2, № 1—2, с. 46—56.

Л ю б и м е н к о В. Н. Курс общей ботаники. Берлин, 1923. 1040 с.

Л ю б и м е н к о В. Н. Фотосинтез и хемосинтез в растительном мире. М.—Л., 1935, с. 5—321. То же: Избр. тр. в 2-х т. Т. I. Киев, 1963, с. 320—597.

Л ю б и щ е в А. А. К логике систематики. — В кн.: Проблемы эволюции. Т. II. М., 1972, с. 45—68.

М а к - Д у г о л л В. Б. Экология растений. М., 1935. 211 с.

М а н о й л е н к о К. В. Развитие эволюционного направления в физиологии растений. Л., 1974, 255 с.

М а н о й л е н к о К. В., Х а х и н а Л. Н. Из истории развития эволюционной теории в Академии наук и вклад академика А. С. Фаминцына. — Журн. общ. биол., 1974, т. 35, № 2, с. 308—314.

М е д н и к о в Б. М. Дарвинизм в XX веке. М., 1975. 223 с.

М е й с т е р Г. К. Критический очерк основных понятий генетики. М.—Л., 1934. 203 с.

М е р е ж к о в с к и й К. С. Материалы к познанию животных пигментов. СПб., 1883. 99 с.

М е р е ж к о в с к и й К. С. К морфологии диатомовых водорослей. Казань. 1903. 429 с.

(М е р е ж к о в с к и й К. С.) M e r e s c h k o w s k y C. Über Natur und Ursprung der Chromatophoren im Pflanzenreiche. — Biol. Centralbl., 1905a, Bd 25, N 18, S. 593—604.

М е р е ж к о в с к и й К. С. По поводу моих работ над эндохромом диатомовых водорослей. — Зап. Ун-та, Казань, 1905б, т. 72, кн. 9, с. 1—44.

М е р е ж к о в с к и й К. С. Законы эндохрома. — Зап. Ун-та, Казань, 1906. т. 73, кн. 2, 3, 5—6, с. 1—176, 177—288, 289—402.

М е р е ж к о в с к и й К. С. Теория двух плазм как основа симбиогенезиса, нового учения о происхождении организмов. Казань, 1909a. 102 с.

М е р е ж к о в с к и й К. С. Конспективный курс споровых растений. Казань, 1909б. 161 с.

М е р е ж к о в с к и й К. С. Конспективный курс общей ботаники. — Зап. Ун-та, Казань, 1910, т. 77. 170 с.

(М е р е ж к о в с к и й К. С.) M é r e j k o v s k y C. La Plante considérée comme un complexe symbiotique. — Bull. Société Sci. Natur., 1920, N 6, p. 17—98.

М о р д в и л к о А. К. Муравьи и тли. — Природа, 1936, № 4, с. 44—53.

М о ш к о в с к и й Ш. Д. Функциональная паразитология. — Мед. паразитология и паразитол. болезни, 1946, т. 15, № 4, с. 26—36.

М о ш к о в с к и й Ш. Д. О природе простейших (Protozoa) и границах протозоологии. — Тр. Ленингр. о-ва естествоиспыт., 1957, т. 73, № 4, с. 129—137.

Н а в а ш и н С. Г. (Реф.) Фаминцын А. О симбиозе водорослей с животными. — В кн.: Обзор ботанической деятельности в России за 1891 г. СПб., 1892, с. 81—86.

Н а в а ш и н С. Г. Принцип преемственности и новые методы в учении о клетке высших растений. — Журн. Рус. бот. о-ва, 1916, т. I, № 1—2, с. 1—38.

Н а в а ш и н С. Г. Пол — фактор органической эволюции. Вологда, 1926. 37 с.

Н а с ы р о в Ю. С. Автономия хлоропластов? — Природа, 1975a, № 11, с. 43—48.

Н а с ы р о в Ю. С. Фотосинтез и генетика хлоропластов. М., 1975б. 144 с.

145

141

Одинцова М. С. ДНК хлоропластов и митохондрий (структура, репликация, физико-химические свойства). — В кн.: Биологическая химия. Т. 10. М., 1976. 178 с.

Одум Е. Экология. М., 1968. 168 с.

Одум Ю. Основы экологии. М., 1975. 740 с.

Окснер А. Н. Флора лишайників України. Т. I. Київ. 495 с.

Окснер А. Н. Определитель лишайников СССР. Вып. 2. Общая часть. Л., 1974. 283 с.

Опарин А. И. Возникновение жизни на Земле. М.—Л., 1941. 267 с.

Опарин А. И. Проблема происхождения жизни. М., 1976. 62 с.

Остроумов С. А. О чем говорит сходство митохондрий, хлоропластов и прокариотов? — Природа, 1973, № 3, с. 21—29.

Павловский Е. Н. Организм как среда обитания. — Природа, 1934, № 1, с. 80—91.

Парнес В. А. Антон де Бари. М., 1972. 190 с.

Пахомова М. В. ДНК водорослей. — В кн.: Строение ДНК и положение организмов в системе. М., 1972, с. 176—195.

Пахомова М. В. Нуклеиновые кислоты синезеленых водорослей. — В кн.: Актуальные проблемы биологии синезеленых водорослей. М., 1974, с. 104—113.

Перфильев Б. В. К учению о симбиозе. — Микробиология, 1914, т. 1, № 3—5, с. 209—224.

Пиневич В. В., Васильева В. Е. Каротиноиды синезеленых водорослей. Значение для понимания эволюции и механизма фотосинтеза. — Вестн. Ленингр. ун-та, 1972, № 21, с. 105—122.

Пиневич В. В., Сойдла Т. Р., Иванова С. Б., Левитина Т. П., Липская А. А. Происхождение фотосинтетических систем эукариотов. (К вопросу о симбиотической гипотезе возникновения хлоропластов). — Цитология, 1974, т. 16, с. 124—136.

Пинус Е. А. Об автономии митохондрий. — В кн.: Проблемы возникновения и сущности жизни. М., 1973, с. 246—250.

Полянский В. И. Избранные главы дарвинизма. Л., 1947. 122 с.

Полянский В. И. [Предисловие к статье А. А. Еленкина «В Восточных Саянах вместе с В. Л. Комаровым»]. — Тр. Ин-та истории естествознания и техники, 1957, № 16, с. 253—334.

Полянский Ю. И., Стрелков А. А. О влиянии фауны инфузорий рубца на рост животных. — Тр. Петергоф. биол. ин-та ЛГУ, 1935, № 13—14, с. 68—87.

Работнов Т. А. О консорциях. — Бюл. МОИП. Отд. биол., 1969, т. 74, № 4, с. 109—116.

Работнов Т. А. Консорция как структурная единица биогеоценоза. — Природа, 1974, № 2, с. 26—35.

Райков Б. Е. Из истории дарвинизма в России. — Тр. Ин-та истории естествознания и техники, 1957, т. 16, с. 3—33.

Робертис Э., Новинский В., Саэс Ф. Биология клетки. М., 1973. 487 с.

Рубин Б. А., Арцеховская Е. В. Биохимия и физиология иммунитета растений. М., 1968. 412 с.

Рудин Д., Уилки Д. Биогенез митохондрий. М., 1970. 156 с.

Руттен М. Происхождение жизни (естественным путем). М., 1973. 411 с.

Руцкий И. А. Светлой памяти профессора Б. М. Козо-Полянского. — Бюл. О-ва естествоиспыт. при Воронеж. ун-те, 1959, № 11, с. 5—11.

Рыжков В. Л. Начатки биологии (история видов и особи). Харьков, 1924. 199 с.

Рыжков В. Л. Новое в биологии. Популярные очерки. Харьков, 1927. 163 с.

Рыжков В. Л. Лабильное состояние генов, генома и цитоплазмы. — Усп. соврем. биол., 1939, т. 11, вып. 2, с. 348—363.

Рыжков В. Л. Проблема эволюции ультравирусов. — Микробиология, 1942, т. 11, вып. 4, с. 149—159.

Р ы ж к о в В. Л. О так называемой инфекционной наследственности у бактерий. — Вопр. вирусологии, 1960, № 5, с. 515—520.

Р ы ж к о в В. Л. Краткий очерк истории изучения вирусов. — Тр. Ин-та истории естествознания и техники, 1961, т. 36, с. 315—325.

Р ы ж к о в В. Л. О симбиозе на молекулярном уровне. — В кн.: Тезисы докладов о природе вирусов к конференции, посвященной 100-летию со дня рождения Д. И. Ивановского. М., 1964, с. 12—15.

Р ы ж к о в В. Л. О симбиозе на молекулярном уровне. — Усп. соврем. биол., 1965, т. 59, вып. 3, с. 385—398.

Р ы ж к о в В. Л. Внутриклеточный симбиоз. — Природа, 1966, № 3, с. 9—17.

Р ы ж к о в В. Л. Что старого в новой генетике? — Вопр. филос., 1976, № 3, с. 89—95.

С а в и ч В. П. Памяти А. А. Еленкина (1873—1942). — Сов. ботаника, 1944, № 1, с. 60—63.

С а м ц е в и ч С. А. Значение клубеньковых бактерий как симбионта бобовых растений. — В кн.: Роль микроорганизмов в питании растений и плодородии почвы. Минск, 1969, с. 4—56.

С а п е г и н А. А. Исследование индивидуальности пластид. Одесса, 1913. 133 с.

С в е т а й л о Э. Н. Некоторые биохимические аспекты проблемы происхождения хлоропластов. — В кн.: Проблемы возникновения и сущности жизни. М., 1973, с. 238—245.

С е д о в а Т. В. Основы цитологии водорослей. Л., 1977. 172 с.

С е н ч е н к о в а Е. М. Андрей Сергеевич Фаминцын. (К 40-летию со дня смерти). — Бот. журн., 1960, т. 45, № 2, с. 309—318.

С е н ч е н к о в а Е. М. Развитие представлений о происхождении и эволюции фотосинтеза растений. — В кн.: Проблемы физиологии растений. Исторические очерки. М., 1969, с. 111—163.

С е н ч е н к о в а Е. М. Развитие представлений об автономии хлоропластов. — В кн.: Из истории биологии. Вып. 4. М., 1973, с. 41—62.

С е н ч е н к о в а Е. М. Открытие хроматографии и Академия наук. — Природа, 1974, № 5, с. 93—101.

С к р я б и н К. И. Симбиоз и паразитизм в природе. Пг., 1923. 205+12 с.

С м и р н о в Л. А. Дважды сделанное открытие. — Бот. журн., 1952, т. 37, № 5—6, с. 896—898.

С т р о г о н о в Б. П. Физиология растений в Академии наук СССР (Возникновение и развитие физиологии растений). — Физиол. раст., 1974, т. 21, № 3, с. 445—454.

С у к а ч е в В. Н., Д ы л и с Н. В. Основы лесной биогеоценологии. М., 1964, 574 с.

С у х о в К. С. Вирусы растений и насекомые-переносчики. М., 1942. 66 с.

С ы т н и к К. М., Г о р д е ц к и й А. В. Физиология растений в свете современной эволюционной теории. — В кн.: Проблемы эволюционной физиологии растений. Л., 1974, с. 12—14.

С э д ж е р Р. Цитологические гены и органеллы. М., 1975. 423 с.

Т а х т а д ж я н А. Л. Б. М. Козо-Полянский. К 60-летию со дня рождения. — Бот. журн., 1950, т. 35, № 4, с. 416—431.

Т а х т а д ж я н А. Л. Четыре царства органического мира. — Природа, 1973, № 2, с. 22—32.

Т а х т а д ж я н А. Л. Растения в системе организмов. — В кн.: Жизнь растений. Т. 1. М., 1974, с. 49—57.

Т а х т а д ж я н А. Л. Система органического мира. — БСЭ. Изд. 3-е. Т. 23. 1976, с. 466—468.

Т е р е х и н Э. С. Лишайники, их происхождение и роль симбиогенеза в эволюции. — Наука и жизнь, 1965, № 1, с. 94—96.

Т и м и р я з е в К. А. Опровергнут ли дарвинизм? — Русская мысль, 1887, кн. 5, № 2, с. 145—180; кн. 6, № 2, с. 1—14. То же: Соч. Т. VII. М., 1939, с. 263—326.

Тимирязев К. А. Растение — сфинкс. — В кн.: Публичные лекции и речи. М., 1888, с. 209—229. То же: Соч. Т. I. М., 1937, с. 294—313.

Тимирязев К. А. Странный образчик научной критики. — Русская мысль, 1889, т. 3, № 2, с. 90—102; то же: Соч. Т. VII. М., 1939, с. 425—443.

Тимирязев К. А. Симбиоз. — Энцикл. словарь «Гранат». Т. 38. 1922, с. 591—596.

Трасс X. X. Лишайники. — Природа, 1973, № 3, с. 10—20.

Трасс X. X. Геоботаника. История и современные тенденции развития. Л., 1976. 252 с.

Фаминцын А. С. Сообщение о работах Швенденера над лишаями. — Тр. СПб. о-ва естествоиспыт., 1870, т. I, с. 39—40.

Фаминцын А. С. Дарвин и его значение в биологии. Речь, читанная на акте в С.-Петербургском университете 8 февраля 1874 г. СПб., 1874. 22 с.

(Фаминцын А.) Famintzin A. Beitrag zur Symbiose von Algen und Thieren. — Mém. Acad. Impér. Sci. St.-Pétersb., Ser., VII, 1889a, Bd 36, N 16, S. 1—36.

Фаминцын А. С. Н. Я. Данилевский и дарвинизм. Опровергнут ли дарвинизм Данилевским? — Вестник Европы, 1889б, т. I, с. 616—643.

Фаминцын А. С. О психической жизни простейших представителей живых существ. — В кн.: Труды VIII съезда русских естествоиспытателей и врачей. Общий отдел. СПб., 1890, с. 32—39.

Фаминцын А. С. О симбиозе водорослей с животными. — Тр. Бот. лаб. Акад. наук, 1891, № 1, с. 1—22.

Фаминцын А. С. О судьбе зерен хлорофилла в семенах и проростках. — Тр. Бот. лаб. Акад. наук, 1893, № 5, с. 1—16.

Фаминцын А. С. Ближайшие задачи биологии. — Вестник Европы, 1894, май, с. 132—153.

Фаминцын А. С. Современное естествознание и психология. — Мир божий, 1898, № 1—7; № 3, с. 167—199.

Фаминцын А. С. Современное естествознание и ближайшая его задача. — Мир божий, 1899, № 12, с. 1—12.

Фаминцын А. С. О роли симбиоза в эволюции организмов. — Зап. Акад. наук. Сер. VIII, физ.-мат. отд-ние, 1907а, т. XX, № 3, с. 1—14.

Фаминцын А. С. О роли симбиоза в эволюции организмов. — Тр. СПб. о-ва естествоиспыт., 1907б, т. 38, вып. 1. Протоколы заседания, № 4. с. 141—143.

Фаминцын А. С. О роли симбиоза в эволюции организмов. — Изв. Акад. наук. Сер. VI, 1912а, т. VI, с. 51—68.

Фаминцын А. С. О роли симбиоза в эволюции организмов. — Изв. Акад. наук. Сер. VI, 1912б, т. VI, № 11, с. 707—714.

Фаминцын А. С. К вопросу о зооспорах у лишайников. — Изв. Акад. наук. Сер. VI, 1914, т. VIII, № 6, с. 429—433.

Фаминцын А. С. О роли симбиоза в эволюции организмов. — Изв. Петрогр. биол. лаборатории, 1916, т. 15, вып. 3—4, с. 3—4.

Фаминцын А. С. Что такое лишайники? — Природа, 1918, апрель—май, с. 266—282.

(Фаминцын А., Баранецкий О.) Famintzin A., Baranetzky I. Zur Entwickelungsgeschichte der Gonidien und Zoosporen — Bildung der Flechten. — Mém. Acad. Impér. Sci. St.-Pétersb., Ser. VII, 1867, Bd XI, N 9, S. 1—6.

Фаминцын А. С., Серк В. Еще о зооспорах у лишайников. Образование зооспор в гонидиях, срощенных с гифами — Изв. Акад. наук, Сер. VI, 1915, т. IX, № 11, с. 1203—1208.

Филиппович И. И., Светайло Э. Н., Алиев К. А. Свойства и особенности основных компонентов белоксинтезирующей системы хлоропластов. — В кн.: Функциональная биохимия клеточных структур. М., 1970, с. 132—142.

148

Филипченко А. А. Экологическая концепция паразитизма и самостоятельность паразитологии как научной дисциплины. — В кн.: Проблемы общей паразитологии. Л.—М., 1937, с. 4—14.

Фокс С., Дозе К. Молекулярная эволюция и возникновение жизни. М., 1975. 374 с.

Хахина Л. Н. Проблема симбиогенеза в работах отечественных ученых. — В кн.: Наука и техника. Вып. 6. Л., 1971, с. 187—190.

Хахина Л. Н. О значении интеграции единиц организменного уровня для эволюционного процесса. — В кн.: Организация и эволюция живого. Л., 1972а, с. 88—92.

Хахина Л. Н. К оценке концепции симбиогенеза в работах Б. М. Козо-Полянского. — В кн.: Наука и техника. Вып. 7. Л., 1972б, с. 30—33.

Хахина Л. Н. К истории учения о симбиогенезе. — В кн.: Из истории биологии. Вып. 4. М., 1973а, с. 63—75.

Хахина Л. Н. Экспериментальные истоки учения о симбиогенезе (работы А. С. Фаминцына). — В кн.: История и теория эволюционного учения. Вып. 1. Л., 1973б, с. 129—141.

Хахина Л. Н. Теоретические истоки взглядов А. С. Фаминцына на роль симбиоза в эволюции организмов. — В кн.: История и теория эволюционного учения. Вып. 1. Л., 1973в, с. 142—149.

Хахина Л. Н. Идея симбиогенеза и место ее в системе эволюционных воззрений А. С. Фаминцына. — В кн.: Наука и техника. Вып. 8. Л., 1973г, с. 181—185.

Хахина Л. Н. К формированию гипотезы симбиогенеза. К. С. Мережковский. — В кн.: История и теория эволюционного учения. Вып. 3. Л., 1975, с. 5—28.

Хейс У. Генетика бактерий и бактериофагов. М., 1965. 555 с.

Хохлов С. С. Симбиогенез: история происхождения клетки и организмов путем симбиоза. Две лекции по эволюционному учению. Саратов, 1977. 46 с.

Хьюитт Л. Влияние бактериофага на изменчивость и эволюцию бактерий. — В кн.: Адаптация у микроорганизмов. М., 1956, с. 422—446.

(Цвет М. С.) Tswett M. Etudes des physiologie cellulaire. Pt. II. Les chloroplastes. — Arch. sci. phys. et natur. Genéve, t. 2, № 7, p. 228—260, 339—348, 467—486, 565—574.

Чайковский Ю. В. Генетическая интеграция клеточных структур как фактор эволюции. — Журн. общ. биол., 1977, т. 38, № 6, с. 823—835.

Черемисинов Н. А. Заседание Воронежского отделения ВБО, посвященное памяти Б. М. Козо-Полянского. — Бот. журн., 1959, т. 44, № 2, с. 275.

Черемисинов Н. А. Козо-Полянский как педагог и ученый-фитопатолог. — Изв. АН СССР. Сер. биол., 1962, № 2, с. 275—282.

Шапошников Г. Х. Живые системы с малой степенью целостности. — Журн. общ. биол., 1975, т. 36, № 3, с. 323—335.

Шемаханова Н. М. Микотрофия древесных пород. М., 1962. 375 с.

Шмальгаузен И. И. Проблемы дарвинизма. Л., 1969. 493 с.

Элементарные процессы генетики. Л., 1973. 255 с.

Яблоков А. В., Юсуфов А. Г. Эволюционное учение. М., 1976. 335 с.

Ярошенко П. Л. Общая биогеография. М., 1975. 188 с.

Ahmadjian V. The Lichen symbiosis: its origin and evolution. — In: Evolutionary Biology. V. 4. New York, 1971, p. 163—182.

Aspects of the biology of symbiosis. Ed. Cheng T. London, 1971. 327 p.

Ball G. Organisms living on and in Protozoa. — In: Researsh in Protozoology, 1969, № 3, p. 565—718.

Baltus R., Brachet I. Presence of deoxyribonucleic acid in the chloroplast of Acetabularia mediterranea. — Biochim. Biophys. Acta, 1963, v. 76, p. 490—492.

Barricelly N. A. Numerical testing of evolution theories. Pt I. Theoretical introduction and basis tests. — Acta Biotheoretica, 1962—1963, v. 16, pt. I—II, p. 70—98.

149

B a r r i c e l l y N. A. Numerical testing of evolution theories. Pt. II. Symbio-
genesis and terrestrial life. — Acta Biotheoretica, 1963, v. 16, pt III—
IV, p. 99—126.
B a r y A. de. Die Erscheinung der Symbiose. Strassburg, 1879. 30 S.
B a t r a S. W. T., B a t r a L. R. The fungus gardens of insects. — Sci. Amer.,
1967, v. 217. N 5, p. 112—120.
B e a l e G. H., J u r a n d A., P r e e r J. R. The classes of endosimbiont
of Paramecium aurelia. — J. Cell Sci., 1969, v. 5, N 1, p. 65—78.
B e n e d e n P. van. Die Schmarotzer des Thierreichs. Leipzig, 1876. 274 S.
B e r n a r d N. L'evolution dans la symbiose. Les Orchidées et leurs champig-
nons commensaux. — Ann. sci. natur., 1909, sér. 9, t. IX, № 1, p. 1—196.
B o n e n L., D o o l i t t l e W. On the prokaryotic nature of red algal chlo-
roplasts. — Proc. Nat. Acad. Sci. USA, 1975, v. 72, N 6, p. 2310—2314.
B u c h n e r P. Endosymbiosis of animals with plant microorganisms. New
York, 1965, p. 3—764.
D u b o s R., K e s s l e r A. Integrative and disintegrative factors in symbiotic
associations. — In: Symbiotic Associations. London, 1963, p. 1—11.
G i b o r A., G r a n i c k S. Plastids and mitochondria inheritable systems. —
Science, 1964, v. 145, N 3632, p. 890—896.
G r a n i c k S., G i b o r A. The DNA of chloroplasts, mitochondria and centrio-
les. — In: Progress in Nucleic Acid Research and Molecular Biology.
V. 6. New York—London, 1967, p. 143—186.
G r e e n P. B. Cinematic observations on the grown and division of chlo-
roplasts in Nitella. — Amer. J. Bot., 1964, v. 51, N 3, p. 334—342.
H a l l W. T., C l a u s G. Ultrastructural studies on the blue-green algae
symbiont in Cyanophora paradoxa Korschikoff. — J. Cell Biol., 1963,
v. 19, N 3, p. 551—560.
H a l l W., C l a u s G. Ultrastructural studies on cyanelles of Glaucocystis
nostochinearum Itzigsohn. — J. Phycology, 1967, v. 3, N 1, p. 37—51.
H a r l e y J. L. The biology of mycorrhiza. London, 1963. 234 p.
H u x l e y J. Evolution. The modern synthesis. London, 1963. 652 p.
I w a m u r a T. Distribution of nucleic acids among subcellular fractions
of Chlorella. — Biochem. Biophys. Acta, 1960, v. 42, N 1, p. 161—163.
L e e R. E. Origin of plastids and the phylogeny of algae. — Nature, 1972,
v. 237, N 5349, p. 44—46.
L i m b a u g h C. Cleaning symbiosis. — Sci. Amer., 1961, v. 205, N 2, p. 42—49.
M a r g u l i s L. Evolutionary criteria in Thallophytes: a radical alternative. —
Science. 1968, v. 161, N 3845, p. 1020—1022.
M a r g u l i s L. Origin of Eukaryotic Cells. New Haven, 1970. 349 p.
M a r g u l i s L. Symbiosis and evolution. — Sci. Amer., 1971a, v. 225, N 2,
p. 48—57.
M a r g u l i s L. Whittaker's five kingdoms of organisms: minor revisions
suggested by considerations of the origin of mitosis. — Evolution (USA),
1971b, v. 25. N 1, p. 242—245.
M a r g u l i s L. Five-kingdom classification and the origin and evolution
of cells. — In: Evolutionary Biology. V. 7. New York—London, 1974,
p. 45—78.
M a r g u l i s L. Symbiotic theory of the origin of eukaryotic organelles:
criteria for proof. — In: Symbiosis. London—New York—Melbourne,
1975, p. 21—34.
M ü h l e t h a l e r K., B e l l P. R. Untersuchungen über die Kontinuität von
Plastiden und Mitochondrien in der Eizelle von Pteridium aquili-
num (L.) Kühn. — Naturwissenschaften, 1962, Bd 49, H. 3, S. 63—64.
N a s s M. M. T. Uptake of isolated chloroplasts by mammalian cells. —
Science, 1969, v. 165, N 3898, p. 1128—1131.
N a s s S. The significance of the structural and functional similarities of bac-
teria and mitochondria. — In: International Review of Cytology. V. 25.
New York—London, 1969, p. 55—117.
N ä g e l i C. Mechanisch-physiologische Theorie der Abstammungslehre. Mün-
chen—Leipzig, 1884. 822 p.

150

O s c h m a n J. L., G r a y P. A study of the fine structure of Convoluta roscoffensis and its endosymbiotic algae. — Transact. Amer. Microscop. Soc., 1965, v. 84, p. 368—375.
P r e e r J. R., P r e e r L. B., J u r a n d A. Kappa and other endosymbionts in Paramecium aurelia. — Bact. Rev., 1974, v. 38, p. 113—163.
R a v e n P. H. A multiple origin for plastids and mitochondria. — Science, 1970, v. 169, N 3946, p. 641—646.
R e i n k e J. Morphologische Abhandlungen. Leipzig, 1873. 122 S.
R e i n k e J. Abhandlungen über Flechten. — Jahrb. wiss. Bot., 1894, S. 495—542; 1895, S. 395—486.
R i s H., P l a u t W. Ultrastructure of DNA — containing areas in the chloroplast of Chlamydomonas. — J. Cell Biol., 1962, v. 13, N 3, p. 381—391.
S a g a n L. On the origin of mitosing cells. — J. Theoret Biol., 1967, v. 14, N 3, p. 225—274.
S c h w e n d e n e r S. Die Algentypen der Flechtengonidien. Basel, 1869. 42 S.
S t a n i e r R. J. Some aspects of the biology of cells and their possible evolutionary significance. — In: Prokaryotic and eukaryotic cells. Symp. soc. gener. microbiol. Cambridge, 1970, p. 1—38.
S t a r r M. P. A generalized scheme of organismic associations. — In: Symbiosis. London—New York—Melbourne, 1975, p. 1—20.
S y m b i o s i s Ed. Henry S. M. V. 1. New York, 1966, 1967.
S y m b i o s i s Ed. Jennings D. H., Lee D. L. London—New York—Melbourne, 1975. 633 p.
S y m b i o t i c Associations. Ed. Nutman P. S., Mosse B. Cambridge, 1963.
T a y l o r D. L. Chloroplasts as symbiotic organelles. — In: International Review of Cytology. V. 27. New York—London, 1970, p. 29—64.
T a y l o r F. J. R. Autogenous theories for the origin of Eukaryotes. — Taxon, 1976, v. 25, N 4, p. 337—390.
T r e n c h R. K., G r e e n e R. W., B y s t r o m B. G. Chloroplasts as functional organelles in animal tissues. — J. Cell Biol., 1969, v. 42, N 2, p. 404—417.
U r b a n e k A. Revolucja naukowa w biologii. Warszawa, 1973. 238 c.
W a l l i n I. E. The symbionticism and the origin of species. London, 1927.
W e r z G., K e l l n e r G. Isolierung und electronenmikroskopische Charakterisierung von Desoxyribonucleinsäure aus Chloroplasten kernloser Acetabularia—Zellen. — Zeitschr. Naturforsch., 1968, N 23b, S. 1018—1019.
W h i t t a k e r R. H. New concepts of the kingdoms of organisms. — Science (Wach.), 1969, v. 163, p. 150—160.
Z a b l e n L. B., K i s s i l M. S., W o e s e C. R., B u e t o n D. E. Phylogenetic origin of the chloroplast and prokaryotic nature of its ribosomal RNA. — Proc. Nat. Acad. Sci. USA, 1975, v. 72, N 6, p. 2418—2422.

Appendix: Ivan E. Wallin and His Theory of Symbionticism

Donna C. Mehos

The American Reaction

Ivan E. Wallin (1883–1969) is virtually unknown to biologists and to historians of biology.[1] His work on cellular evolution in the 1920s was summarily rejected by leading figures in early twentieth-century American biology. At that time, biology had recently changed dramatically in practice and in professionalization.[2] Researchers embraced new experimental analytical techniques as they turned away from the descriptive biology that had characterized most of the nineteenth century. With the "new biology" emerged social rules and institutional allegiances that determined who would be accepted into this redefined profession. Wallin's exclusion from it illustrates how scientists and scientific discourse influence the careers of other, particularly unorthodox, researchers.

Wallin held the unconventional view that intracellular symbiosis played a crucial role in evolution—based, in part, on his belief that some cytoplasmic components evolved from symbiotic bacteria. He was well aware that most biologists viewed symbioses as curiosities. When he discussed his work, Wallin appealed to established scientists by saying "the conception

that all living cells are loaded with bacteria is so startling and antagonistic to our orthodox notions of the cell, that without further analysis and reflection the conception appears absurd."[3] The idea that normal cells were loaded with bacteria not only was unacceptable to Wallin's colleagues, but also it carried implications for many areas of biological inquiry. Wallin saw what he called *symbionticism* as the vehicle for revolutionary conceptions in heredity, development, and cell structure.[4] Working between 1915 and 1927, Wallin performed experimental work that, coupled with his extensive knowledge of scientific literature, led him to propose symbiosis not only as a mechanism for the origin of mitochondria but also for the origin of species in general. Biologists powerful within the field disagreed. Defeated by his opponents, Wallin withdrew from scientific discourse entirely after 1927. Why was he an outsider whose work was rejected? Wallin's position as professor of anatomy at the Medical School of the University of Colorado placed him in geographical as well as intellectual isolation. Scientists working jointly on similar problems were then concentrated on the East Coast. The Marine

149

Ivan E. Wallin (1883–1969). Photograph courtesy of the Denison Memorial Library,
University of Colorado

Biological Laboratory in Woods Hole, Johns Hopkins, Cornell, Princeton, and particularly Columbia—institutions that had developed specific cooperative research programs—were the centers for experimental embryology, cytology, development, and heredity studies. Wallin had no professional relations with these institutions. Furthermore, at a time when relations between mentor and student were becoming important in perpetuating newly established research traditions, Wallin had no mentor and no graduate students.

The story of Wallin raises many questions. During an exciting period in the history of American biology, Wallin was a clever biologist who never fit into the mainstream. He was a man alone who attempted a theoretical synthesis that linked many fields in biology and answered at least one of the most puzzling questions in evolutionary theory: What is the source of evolutionary novelty, of hereditary variation that is selected in nature? His idiosyncratic ideas and irreproducible experimental work provided a basis for criticism by the scientific authorities, primarily cytologists, who judged the merit of biological work on its experimental success. Wallin's rejected theory and the controversy that surrounded it, however, also reveal a tightly knit social structure in which Wallin was not included.

Historical Background
of Cytology

In the mid-nineteenth century, physiologists such as Claude Bernard and Carl Ludwig established a reductionist experimental tradition that spread to other fields in research biology, especially embryology. Wilhelm Roux, Wilhelm His, and other experimental embryologists of the late 1880s changed biological practices not only by forming hypotheses and testing them, but also by emphasizing the role of physicochemical laws in living systems. Scientists in most fields of biology began to use the experimental approach borrowed from the physical sciences. New experimental biologists considered unsupported speculation the bane of biologists as successful experimental researchers attempted to raise the status of biology to that of a "true" scientific discipline. The rejection of Wallin's work was entirely consistent with the sentiment to purge biology of its broad undisciplined natural philosophy and to replace it with precise statements derived strictly from experimental "facts."

Biologists involved in the development of cytology as an independent discipline transformed and redefined biological studies between 1880 and 1930. The influential cell biologist E. B. Wilson's delineation of three eras in the history of cytology is useful here.[5] The three periods began approximately in the years 1840, 1870, and 1900. In the first period researchers developed cell theory. Morphological cytology emerged from histology as a branch of medical anatomy, separating from it later. As biologists established that cells did not arise de novo, they came to see heredity as the genetic continuity of single cells. Embryologists who studied cleavage realized that mitotic cell division maintained genetic continuity and that the germ cells reestablished the cleaving embryo with each generation, thereby providing the physical basis of heredity. The search to understand just what substance germ cells transmitted led to Wilson's second era.

Cytology developed as a discipline truly separate from histology as a result of embryology research, especially fertilization studies. Oscar Hertwig and Eduard Strasburger discovered fertilization in animals and in plants, respectively; these observations were made possible in part by the development of substage condenser lenses on compound optical microscopes. They saw that maternal and paternal nuclei fused or became closely associated, the fusion product becoming the first embryonic nucleus. Later, E. van Beneden discovered that parents each contribute one half to the offspring's chromosomes. After these findings in the 1870s and 1880s, cytology was established as a discipline with its own set of problems. Although the cytoplasm and its inclusions were also under study at the time, research on the nucleus provided far more dramatic results, with most cytologists focusing on it, and more specifically on the chromosomes within it, because they believed that the study of chromosomes would reveal the secrets of heredity.

Cytology expanded greatly in the decade of 1890. Observations of cells confirmed the generality of the mitotic process and aided investigations into mechanisms of heredity and evolution. Zoologists and botanists studied the nuclei of tissues and deciphered the cyclic behavior of chromosomes. Cytoplasm, which had not yet been conceptualized, was soon considered as living matter, and many ideas of its structure, function, and components arose. Observers of the cytoplasm noted small moving "dots" with changing appearances and queried the possible function of these dots. Failure to resolve their biochemistry and structure precluded both assigning them a function and identifying them in all of their guises. Researchers debated the nature of these mitochondria. The identity and meaning of the smaller, less distinguishable mitochondria had always been more equivocal than the similar, larger, and more conspicuous chloroplasts of algal and plant cells. Indeed, there were even attempts to classify these various cell components as living or lifeless. Most of these findings proved inconclusive. In the absence of criteria to define *living* and *nonliving*, the significance of cell inclusions remained disputable and provoked much heated debate among biologists. It was into this arena of debate that Ivan E. Wallin boldly entered.

Wallin's Scientific Work

Wallin studied the lamprey *Ammocetes* for his dissertation in anatomy at New York University. His research focused on histological studies of tissue development, differentiation, and morphology and reflected Wallin's early interest in evolution.[6] At Colorado, he became a dedicated classroom teacher (telephone interviews with Leo Massopust and Carl Weingartner, 16 November 1983; and with John T. Willson, 17 November 1983) but not a dedicated anatomist. After receiving his D.Sc. in 1915, he never did anatomical research again, and there is no obvious connection between his training and academic affiliation and the impetus for his symbiotic theory of evolution.

Wallin's mitochondrial work was first reported in the *American Journal of Anatomy* between 1922 and 1925. In a series of nine papers entitled "On the Nature of Mitochondria," Wallin described his performance of experimental techniques commonly used by cytologists to distinguish bacteria from the morphologically identical mitochondria. These standardized procedures, he reported, did not in fact work. Other bacteriologists and cytologists, particularly E. V. Cowdry of the Rockefeller Institute, believed that mitochondria reacted differently from bacteria when exposed to the same staining, temperature, and chemical treatments, even though mitochondria and bacteria were microscopically indistinguishable. Wallin performed similar experiments but reached different conclusions. Because bacteria and mitochondria reacted similarly to treatments in Wallin's laboratory, he

asserted that mitochondria were bacteria. Their similarities, however, led him to systematic research to compare them. Wallin's experiments supported the evolutionary theory he tested, a theory built long before he had the experimental results to support it.

To defend experimentally his hypothesis that mitochondria were bacterial symbionts, Wallin needed to isolate mitochondria outside the cell and to maintain them as bacterial cultures. The French medical researcher Paul Portier had already claimed to have cultured mitochondria outside of cells,[7] but Wallin criticized Portier's technique when he accused him of culturing bacterial contaminants. Wallin had performed many experiments in which he believed he had obtained pure cultures of mitochondria.[8] In his published reports, he described in rigorous detail his methods for culturing mitochondria. He was defensive— probably because he expected criticism. What he did not anticipate was that his methods would fail him. After the University of Colorado Medical School moved from Boulder to Denver in 1924, Wallin was unable to reproduce his own experiments.[9] Thus the failure of his most important experiments, in principle confirming that mitochondria were symbiotic bacteria, added greatly to the fuel of his critics' fire.

Wallin published his most significant theoretical paper, "Symbionticism and Prototaxis, Two Fundamental Biological Principles," in 1923, just one year after he reported his early mitochondria experiments.[10] His theoretical evolutionary model of the origin of mitochondria and their biological activity was thoroughly developed by this time. In it, he introduced the word *symbionticism*, defining it as symbiosis on the cellular level. Although other cytologists debated whether or not mitochondria were merely lifeless passive storage elements of cells, Wallin never doubted their biological activity.[11]

In his paper, Wallin filled a crucial void in Darwinian theory with his most original idea that symbionticism "insures the origin of the species." Darwin and his followers had not provided the physical basis of variation in nature that would explain how new species arose. Evolutionists contemporary with Wallin saw natural selection as a force that influenced and sustained changes only in populations of organisms. Wallin's symbiotic theory, however, provided not only a mechanism for the origin of new species but also of the entities: symbiotic aggregates, on which natural selection could act. Through the development of symbiotic relationships, the different species that group together create, in essence, a new organism—new by virtue of its modified physiology and its subsequently altered ability to survive in its environment. Wallin was not shy when he proclaimed: " 'microsymbiosis' . . . offer[s] the solution to the problem of the origin of species."[12]

In his only monograph, *Symbionti-cism and the Origin of Species*, published in 1927, Wallin elaborated on his synthetic theory. To support many of his claims, he drew on the work of the founders of modern genetics, such as W. Bateson, T. H. Morgan, W. E. Castle, C. B. Bridges, A. H. Sturtevant, and others who demonstrated that chromosomes in the nucleus were responsible for the transmission of heredity. These geneticists thought that the rearrangements and mutations of genes might provide a mechanism for the origin of species. Wallin, however, went much further in suggesting that gene transfers occurred from symbionts to the nuclei of their hosts.[13] This striking example illustrates his ability to incorporate findings from various sources as supportive evidence. Wallin also disagreed with the belief common among geneticists that genes were preformed or limited: Different organisms had the same genes arranged in different sequences. Rather, he believed that new species originated by the addition or exchange of genetic material.[14] He therefore borrowed the genetic phenomenon of crossing over—the physical exchange of genes between two chromosomes—to explain the relation of the symbiotic mitochondrion to its host. Referring to current genetic and evolutionary studies, Wallin explained that

it appears possible for a part or all of the chromatin of a microsymbiont to be given up to the nucleus of the host cell and germ plasm when symbiosis develops, leaving the remains of the microsymbiont in the cytoplasm somewhat in the nature of the plast. The part of the chromatin removed from the microsymbiont may not be essential for its growth and reproduction, but it is conceivable that the specific activity of the mitochondria is maintained only under the influence of this chromatin. This influence may be exerted "at long range" from the nucleus.[15]

Wallin derived his explanations from chromosomal research done by the Morgan school and placed his intracellular symbiosis theory in the framework of contemporary genetics. Perhaps he anticipated less controversy by presenting symbionticism in this genetics context, which was the foremost concern of biologists at the time. Wallin's eclectic explanation of the nature of mitochondria did not agree with any contemporary theory, yet he justified what his critics considered wild speculations by saying that all gene concepts were at least as speculative as his own.

Wallin's Critics and Their Rejection of Speculative Biology

Late nineteenth- and early twentieth-century biologists created their discipline based on analytical experimental methods. They purged biology of what they called speculation, although many simply replaced this word with "theory." Presumably, acceptable theory matched more closely their empirical findings than did speculation in an earlier, nonexperimental era of biology. The transition in biology from a speculative, descriptive enterprise to a laboratory science has

been the subject of debate among historians of biology. The disagreement focuses on the nature of this change: Was it gradual or was it the resolution of a dichotomy that had kept empiricism and speculation apart?[16] As with many competing historical models, coherent arguments can be made for both sides. Wallin's rejected work combined experimentation and speculation or theorizing. His experience tells us little about the changing nature of conventions in biological research. More relevant to Wallin's case, however, is the argument employed by J. Sapp in his study of cytoplasmic inheritance research, in which he argues that the rise and success of nucleocentric studies of heredity were the result of a struggle for scientific authority among biologists who wanted to control the field.[17] Certainly, the controversy that surrounded Wallin's work can be seen in that light.

In the late 1920s, Wallin's symbionticism received much attention. Wilson discussed Wallin's theory of symbiosis in evolution in *The Cell in Development and Heredity*; an unfavorable book review appeared in *Nature;*[18] and the *New York Herald American* published a popularized review of Wallin's work.[19] The symbiotic theory of the nature of mitochondria provoked strong reactions, the most damaging of which came from Cowdry.[20] Wallin and Cowdry were bitter enemies (Leo Massopust and Salmon Halpern, telephone conversations with author, 16 November 1983). Their battles were published in the *American Journal of Anatomy* and the *Journal of Experimental Medicine* as well as witnessed at professional meetings of the American Association of Anatomists.[21] The Cowdry controversy, coupled with Wallin's lack of acceptable experimental data, left Wallin completely defeated. After 1927, he did not participate in scientific research again.

Although Cowdry criticized both Wallin's technique and his theory, he apparently objected most strongly to the act of theorizing. Cowdry advocated the position taken by some experimental biologists who viewed natural history as unbridled speculation and who saw most theory as such speculation. For Cowdry, all theory wandered too far from detailed laboratory observations. He wanted the cytological experimental approach to provide quantitative mathematical descriptions of the chemistry of the cell. Cowdry said Wallin did not have enough evidence to support his unorthodox claim that mitochondria were symbiotic bacteria; Wallin said he was in the early experimental stages, and even though he did not yet have enough evidence to convince everyone of this, the evidence for the conventional, cytoplasmic origin of mitochondria was even less convincing.

In his chapter in *General Cytology,* the textbook he edited, Cowdry elucidated his ideas about mitochondria and about theoretical issues in science.[22] In agreement with Wallin, he defended the existence and importance

of mitochondria. Both felt that biologists underestimated the importance of these cellular components. Cowdry devoted quite some space to a criticism of symbionticism, citing Wallin, Portier, and his own previously published critiques of their work. He claimed that there was little experimental evidence to support the symbiosis hypothesis, pointing to his own studies in which he purportedly differentiated between mitochondria and bacteria within the same cell. In contrast to "investigators familiar with mitochondria," only naive biologists could believe mitochondria to be bacteria.[23] In spite of admitting striking morphological similarities between bacteria and mitochondria, Cowdry completely rejected Wallin's symbiotic hypothesis. Furthermore, he argued that neither his nor Wallin's physiological view of mitochondria was fully supported or proved by data. Cowdry openly accepted the dualism of two working hypotheses regarding mitochondrial function: They are passively or actively involved in cell function. However, Cowdry would not tolerate two hypotheses regarding the origin of mitochondria.

Because cytologists simply did not have the laboratory techniques necessary to support their claims, many biologists doubted all such cytological observations. Although Wallin's critics accused him of a lack of definitive results with dyes, chemical reactions, and culture experiments, they, too, faced the same technical limitations. Light microscopic study could not definitively differentiate bacteria from mitochondria. This suggests another motivation for such vehement reactions. Wallin's critics simply did not like he was saying, and they used the most powerful weapon available in that period of biology—methodological criticism.

Cowdry's writings on mitochondria illustrate his position against theoretical biology or speculation, as he called it. In commenting on the contemporary views of mitochondria as sites of cell respiration, he explained that although the idea "has been well received by cytologists and serves as a useful and convenient working hypothesis, it is still only a theory and must be regarded as such."[24] Later he discussed the more generally accepted association of mitochondria and chemical changes in the cell, and reported that it "is happily not a theory, but a fact established by direct and repeated observation."[25]

Why did Cowdry choose to oppose Wallin so strongly? He might have resented what he perceived to be Wallin's intrusion into cytology, the field in which Cowdry was so influential; Wallin had never become part of the large community that worked on cell studies. Furthermore, Cowdry considered evolution to be an untestable hypothesis. Cowdry probably objected to the great significance Wallin attributed to cooperation in nature. Those who knew only a little about Darwinian theory knew of the great significance Darwinists attributed to competition.

At the same time that Cowdry attacked Wallin, Wilson discussed symbionticism. Wilson was neither an opponent nor a supporter of the theory; he was skeptical of all claims about intracellular structures because of the technical limitations in their study. In *The Cell in Development and Heredity,* Wilson described many possible evolutionary origins for various structural components of the cell. He referred to the unverifiability and therefore the apparent worthlessness of a hypothesis of the symbiotic theory of evolution and of cell components. Although it was "a purely speculative construction . . . it seems to the author to offer possibilities concerning the early evolution of the cell that are worth considering."[26] Wilson described the symbiotic theory of Merezhkovsky, who suggested that the origin of the nucleus arose from two symbiotic bacteria, as "an entertaining fantasy."[27] Despite his criticism and strong reservations, Wilson wrote that "to many, no doubt, such speculations may appear too fantastic for mention in polite biological society; nevertheless it is within the range of possibility that they may some day call for more serious consideration."[28]

Wilson was probably more amenable to symbiotic theories than Cowdry because he was both realistic about cytological techniques and deeply interested in evolution. In general, he developed his own theoretical considerations but rarely discussed theory in association with experimental work. He thus supported the dichotomy contemporary biologists drew between experimentation and theory.

The opposition Wallin faced illustrates more general trends of the time. Biological theory was suspect in the 1920s because biologists were determined to vindicate their field as a pure, mature science. In order to build a stronger foundation for biology in the world of science, biologists rejected descriptive and comparative research aimed at broad problems and turned instead to experimental techniques and to narrower questions. Experimentalists generally perceived the work of many nineteenth-century biologists as wild speculation—wild because the theories explained much more than the evidence could support. Some early twentieth-century biologists rejected evolutionary theories entirely because they could not be proved in a laboratory. However, the conscious elimination of speculation was, in part, a rhetorical campaign. Modified to fit the needs of the new experimentalists, speculation reemerged as biological theory and maintained an important role in biological practice. Unlike Cowdry, most experimentalists never entirely rejected theory. Like Wilson, they believed there was a proper context for the discussion of theory in biological research. The reception of Wallin's work represents their reaction against an earlier style of theoretical evolutionary thinking.

Wallin's Isolation

Ivan Wallin produced and presented his work in the absence of any peers,

colleagues, or students—an isolation unheard of in twentieth-century science. Moreover, he apparently had no communication with scientists who might have become his colleagues. There were at least three biologists publishing their symbiosis studies at the same time. In France, Paul Portier claimed he had grown mitochondria in culture and argued for their bacterial origin.[29] Portier thus made exactly the same claims as Wallin. Although well respected for his contributions to medicine, Portier received such vehement criticism of his symbiosis work in A. Lumière's book *Le Myth des Symbiotes* that, much like Wallin, Portier ceased work on cell evolution.[30] Surely Portier is someone with whom Wallin should have been in written communication, but rather, failing to see Portier as a potential close colleague and ally in his battle, Wallin chose to criticize Portier's experimental technique. Another scientist whom Wallin read but did not know was the German Paul Buchner, perhaps the master of symbiosis research. Buchner was the first scientist to convince his colleagues that symbiosis existed as an important widespread biological phenomenon in plants and in animals.[31] The Russian, Konstantine S. Merezhkovsky, also a potentially important ally for Wallin, advanced the theory that plastids—the photosynthesizing organelles in algal and plant cells—originated from photosynthetic symbiotic microorganisms.[32]

Wallin cited the work of Portier, Buchner, and Merezhkovsky but did not correspond with them. In neglecting to form scientific collegialities with those deeply interested and knowledgeable about symbiosis and evolution, Wallin neglected important interchanges with people who might have sympathized with and encouraged him. Yet he worked well with people and had the sense of politics required by all good administrators, evident in his long administrative career at the Medical School of the University of Colorado.[33] His success with his Colorado colleagues is inconsistent with his avoidance of relations with scientific peers. Communication and cooperation between peers in scientific endeavor are essential qualities of twentieth-century science.

The Rise of Nucleocentric
Experimental Biology

It is ironic that Wallin's work was both too early and too late. In the last three decades of the nineteenth century, his work might have been completely acceptable. The earliest cytologists looked at many different organisms, cells, and their parts with so many different concepts that Wallin would not have been unconventional among them. But Wallin's work appeared as the new experimental biologists attributed all significant hereditary roles to the nucleus and its chromosomes; the free and easy period of cell studies ended, and a limited experimental cytology that focused on the nucleus triumphed. Furthermore, the proponents of nucleocentric experimental cytology who waged crusades

in support of their approach arose as the victors who would control much of twentieth-century biological research.

The successful linkage of Mendelian phenomena to experimental results reinforced the opinions of cytologists that experimentalism was the only valid approach. Biologists eventually became enthusiastic about the nucleus as the controlling center of the cell because chromosome studies had confirmed Mendelian predictions—phenomena well known to breeders and agricultural researchers but not well integrated into the formal structures of science.[34] Despite early twentieth-century investigations that linked Mendelian inheritance to experimental data, biologists—particularly those not interested in the practical problems of agricultural research—did not readily accept Mendelism as a general phenomenon in inheritance. Because Mendel's explanation depended on hypothetical hereditary particles, his work was received by professional cytologists and other microscopists with skepticism. What were these particles that nobody could see? The unseen morphological unit with an unexplained physiological function was not easily accepted in this period when scientists rejected theories because of inconclusive experimental evidence.

During the decade of 1910, scientists not only accepted Mendelism but also extended it into cytology and created a unified approach to the behavior of the nucleus and its contents. Thomas H. Morgan and his coworkers identified genes as the physical and mechanical basis of Mendelian inheritance by using successful experimental methods. The gene theory brought together cytological investigations, breeding results, and the hypothetical hereditary particles proposed by Mendel. This work strengthened the arguments made by scientists like Cowdry in support of narrowly conceived experimental biology.

Because the success of Morgan's group was overwhelming, there was little room for the recognition of cytologists whose work did not focus on the nucleus nor on the chromosomal theory of heredity. The chromosomal theory and work on *Drosophila* provided solid evidence for the promise of experimental methods in evolutionary research. Just as geneticists identified chromosomes as the material basis of heredity, Wallin claimed to have solved the problem of the mechanism for evolutionary change by looking outside the nucleus at mitochondria. In these early days, celebrating the triumphs of chromosome mapping and rearrangements, his theory that mitochondria played an important role in evolution—in part based on totally questionable experimental evidence—simply was not acceptable.

In the history of cytological research and its relations to heredity and evolution, the focus of research narrowed to the nucleus in general and to the chromosomes in particular. There are many reasons behind the attraction of the nucleus for cell studies. Cytologists were limited by the quality of their microscopes; nuclei and some chromosomes were more visible than cytoplasmic structures. Because rivaling ideas about mitochondria could not be resolved experimentally, dogmatic claims of researchers—including those of Cowdry and Wallin—were not more reliable than those from preceding decades. Therefore biologists committed to experimentalism concluded that the nucleus was the key to the difficult lock of heredity. This view of the nucleus prevails today, an overemphasis from which biology has not yet recovered.

Wallin was one of the few individuals of his time to work on symbiosis and evolution on the cellular level. His views clashed with those of mainstream biologists who were preoccupied with chromosome experiments. Wallin took the single idea that mitochondria are symbiotic bacteria and expanded it into a theory of the origin of species. He was not always self-consistent, and he used methods different from those of his contemporaries. His critics did not believe his claims to have cultured mitochondria, and they disliked his sweeping, and experimentally unjustified, statements.

Despite opposition, Wallin carried the thread of symbiosis research through periods when it was not fashionable. The enormous social obstacles he faced occurred because he contradicted strong trends in biological research. He created an unconventional answer to the major question riddling those interested in evolution: How did variation arise? Biologists concerned with the confirmation of Darwin's theory of natural selection expected the answer to come from studies of chromosomes; Wallin's solution conflicted with too many of their assumptions. Despite Wallin's synthetic evolutionary theory, which included a physical basis for the origin of variations, it is no surprise that his work was rejected by the scientific community.

Although his major thesis has been vindicated by current research,[35] the rejection of Wallin's theory was based on more than what was considered right or wrong to his contemporaries. Wallin, whose case is not unique in the history of science, broke the rules of those institutions and traditions that guide science. His transgressions and eventual defeat illustrate the predicaments that unconventional scientists have faced. As an independent investigator, Wallin was silenced by those who regulated the conditions under which biologists could be successful. In his own words, he was an "investigator . . . who . . . dared to penetrate the 'theoretical fortifications' . . . built around mitochondria."[36] Although he did dare to penetrate the fortifications built around biological research, he remained on the outside.

Acknowledgments

I would like to thank the many people who helped me to complete this work. In particular, Diana E. Long and Lynn Margulis read the earliest drafts of it when I was a graduate student at Boston University. Along with Virginia P. Dawson, Olga Amsterdamska, and Anja Hiddinga, they offered invaluable suggestions for revision. Frank B. Rogers and William J. Haas provided information about Wallin, as did the staff of the Archives of the Denison Memorial Library of the University of Colorado, which also kindly provided a photograph of Wallin.

This work was supported in part by a Watkins Visiting Professorship at Wichita State University, through National Aeronautics and Space Administration Grant NGR 22–004–025 to Dr. Lynn Margulis, and by the NASA Lewis Research Center.

Notes to Appendix

1. See Jan Sapp, "Symbiosis in Evolution: An Origin Story," *Endocytobiosis and Cell Research* 7(1990): 5–36, and idem., "Living Together: Symbiosis and Cytoplasmic Inheritance," in *Symbiosis as a Source of Evolutionary Innovation: Speciation and Morphogenesis,* ed. L. Margulis and R. Fester (Cambridge: MIT Press, 1991), 15–25, for brief discussions of Wallin's ideas.

2. Historians have shown much interest in the rise of experimental biology and genetics in America as well as in the changing professional strategies and structure of biology. See, for example, "Special Section on American Morphology at the Turn of the Century," *Journal of the History of Biology* 14 (1981): 83–191; Garland Allen, "Naturalists and Experimentalists: The Genotype and the Phenotype," *Studies*

in the History of Biology 3 (1979); idem., *Life Science in the Twentieth Century* (New York: Cambridge University Press, 1978); Jane Maienschein, ed., *Defining Biology: Lectures from the 1890s* (Cambridge: Harvard University Press, 1986); Ronald Rainger, Keith R. Benson, and Jane Maienschein, eds., *The American Development of Biology* (Philadelphia: University of Pennsylvania Press, 1988); Philip Pauly, "The Appearance of Academic Biology in Late Nineteenth-Century America," *Journal of the History of Biology* 17 (1984): 369–97; idem., *Controlling Life: Jacques Loeb and the Engineering Ideal in Biology* (New York: Oxford University Press, 1987); Jan Sapp, *Beyond the Gene: Cytoplasmic Inheritance and the Struggle for Authority in Genetics* (New York: Oxford University Press, 1987).

3. Ivan E. Wallin, "On the Nature of Mitochondria." Pt. 9, "Demonstration of the Bacterial Nature of Mitochondria." *American Journal of Anatomy* 36 (1925): 139.

4. Wallin, *Symbionticism and the Origin of Species* (Baltimore: Williams and Wilkins, 1927), p. viii.

5. E. B. Wilson, Introduction to *General Cytology,* ed. E. V. Cowdry, (Chicago: University of Chicago Press, 1924), 1–12; idem., *The Cell in Development and Heredity,* 3d edition (New York: Macmillan, 1925), 1–20.

6. Wallin, "Thymus-like Structures in *Ammocetes,*" *American Journal of Anatomy,* 22 (1917); idem, "On the Branchial Epithelium of *Ammocetes,*" *Anatomical Record* 14 (1918).

7. Paul Portier, *Les Symbiotes* (Paris: Masson et Cie, 1918).

8. Wallin, "On the Nature of Mitochondria, VIII. Further Experiments in the Cultivation of Mitochondria," *American Journal of Anatomy* 35 (1925); "On the Nature of Mitochondria, IX. Demonstrations of the Bacterial Nature of Mitochondria," *American Journal of Anatomy* 36 (1925).

9. Ironically, Wallin probably cultured bacterial contaminants in his first "successful" experiments.

10. Wallin, "Symbionticism and Prototaxis: Two Fundamental Biological Principles," *Anatomical Record* 26 (1923).

11. In his book *Symbionticism* (see n. 4), Wallin explicated his ideas of the various roles mitochondria play in cell function, such as urea utilization, cell division, differentiation, mitosis, and cancer.

12. Wallin, *Symbionticism,* 7.

13. Ibid., 123.

14. Ibid., 121–22.

15. Ibid., 127.

16. "Special Section on American Morphology"; and Garland Allen, *Life Science in the Twentieth Century* (Cambridge: Cambridge University Press, 1978).

17. Sapp, *Beyond the Gene.*

18. J. Bronte Gatenby, "Nature of Cytoplasmic Inclusion," review of *Symbionticism and the Origin of Species,* by Ivan E. Wallin, *Nature* 12 (1928): 164–65.

19. E. E. Free, "New Theory Hails Tiny Bacteria within Body, as Masters of Man," *New York Herald American,* 21 Nov. 1926.

20. E. V. Cowdry and Peter Olitsky, "Differences between Mitochondria and Bacteria," *Journal of Experimental*

Medicine 36 (1922): 521–33; idem., "The Independence of Mitochondria and the *Bacillus radicicola* in Root Nodules," *American Journal of Anatomy* 31 (1923): 339–41.

21. See n. 3; Ivan E. Wallin, "The Mitochondria Problem," *American Naturalist* 57 (1923); Ivan E. Wallin, "Observation on Mitochondrial Growth in Artificial Culture Media," *Proceedings of the Society for Experimental Biology and Medicine* 25 (1928): 371–72.

22. E. V. Cowdry, "Mitochondria, Golgi Apparatus, and Chromidial Substance," in *General Cytology,* ed. E. V. Cowdry (Chicago: University of Chicago Press, 1924), 311–82.

23. Ibid., 322.

24. Ibid., 325.

25. Ibid., 333.

26. E. B. Wilson, *The Cell in Development and Heredity,* 3d edition (New York: Macmillan, 1925), 738.

27. Ibid.

28. Ibid., 739.

29. See n. 7.

30. Antoine Lumière, *Le Myth des Symbiotes* (Paris: Masson et Cie, 1919).

31. See, e.g., Paul Buchner, *Endosymbiosis of Animals with Plantlike Microorganisms* (New York: Wiley Interscience, 1965). In this work, however, Buchner cites Wallin in the section "Wrong Paths in Symbiosis Research," 70–71.

32. See, e.g., Konstantine Merezhkovsky, "Über Natur und Ursprung der Chromatophoren im Pflanzenreiche," *Biologische Centralblatt* 25 (1905), and "Theorie der Zwei Plasmaarten als Grundlage der Symbiogenesis, einer Neuen Lehre von der Entstellung der Organismen," *Biologische Centralblatt* 30 (1910).

33. Wallin held his position as head of the Department of Anatomy from the time of his appointment in 1918 until he retired in 1951. In the first years he worked closely with the dean, Maurice Riece, to develop the Anatomy Department. More importantly, he was involved in the reorganization of the Medical School during the period when the Rockefeller Foundation awarded it several large grants between 1920 and 1928. Outside of the university, Wallin and Dean Riece established the Colorado State Anatomical Board, which assumed responsibility for the distribution of unclaimed human cadavers to the Medical School "for the promotion of medical science" (Anatomical Law of the State of Colorado, 1927).

34. See Diane B. Paul and Barbara A. Kimmelman, "Mendel in America: Theory and Practice, 1900–1919," in *The American Development of Biology,* ed. Rainger, Benson, and Maienschein, (n. 2) for a discussion of the breeders and academic researchers working on agricultural problems who accepted Mendelism before biologists in arts and sciences universities did; see also Allen, *Life Science,* 50.

35. See, e.g., Lynn Margulis, *Symbiosis in Cell Evolution.* 2d ed. (New York: W. H. Freeman, 1992).

36. See n. 10, "Symbiosis and Prototaxis," 72. Wallin was describing Paul Portier.

INDEX

cellulose, 11, 95

centriole-kinetosomes, 74

centrioles, xvi, xx, 73, 114

centromeres, 114

centrosomes, xx

Cephalopoda, 76

Chaetoproteus, 114

Chatton, E., 70n

chemistry, viii

chemoautotrophy, 113

chemosynthesis, 113

chloramphenicol, 100

chlorellae, 26, 72, 73

chlorobacteria, 71

Chlorobium, 71, 85

Chlorochromatium, 71, 85

Chlorococcum, 55

chlorophyll, 38, 39, 41, 66, 73

chlorophytes, 21n, 97

chloroplasts: and algae, 40–41, 46; cyanelles and, 96; cyanobacteria and, 103; evolutionary success of, xxii; Famintsyn, xix, 20, 26–28, 34, 82, 92; genetic continuity of, 49, 98–102; A. G. Genkel', 82; intracellular reproduction of, 38, 84; lack of, in cytodes, 70; lack of, in mycoids, 45; lichens as models, 21; Merezhkovsky, xix, 34; and mitochondria, 152; origin of, xii, 47, 60–61, 73, 81, 93; photosynthesis, origin of, 85; *Trebouxia,* 55

chondriosomes. *See* mitochondria

chromatin, 154–55

chromatophores: and cytoplasm, 45; definition, xxix; of diatoms, 36–37; leucoplasts, development from, 27; Merezhkovsky, 38–41, 48; from mycoids, 42; origin of, 34, 49, 84; zoochlorellae of ciliates, 26

chromosomes, 48, 107, 114, 152, 154, 159–61

Chroococcus, 97

cilia, xvii, 112

ciliates, 11, 26, 40, 72

Cladonia, 21–22

Clostridium, 71

coccus, 91

Codium, 98

Coelenterata, 11, 72

cohabitation, 7

colicin, 109

Collema, 13, 22

colonies, 70

commensalism, 7, 54, 68

community, 66

Comparative Organollography of Cytoplasm, The, 93

compartmentalization, 116–17

competition, 157

complexification, 1, 2, 6, 43, 73, 92, 120

Concepts of Symbiogenesis, xii, xviii

Concise Course on Cryptogamic Plants, A, 48

Concise Course on General Botany, A, 48

conjugation, 109, 111

consortia, 8–9, 10, 54, 59, 68–69, 70, 74

continental drift, xv

convergence, 78

Convoluta, 72

cooperation, 157

coral, 36

corn, 107

Correns, K., 101

Cowdry, E. V., 153, 155–58, 160

crossing over, 154

crustose lichens, 12–14

cryptogams, 48

cryptomonads, 72, 97

cyanelles, 41, 60–61, 96–97

cyanobacteria (blue-greens): as cytodes, 70; descendants of bacteria, 43; diversity of, 44; DNA comparisons, 105; Elenkin, 60–61; encyclopedia entry, xi; endocyanelles, 96–97; extracellular symbionts, 11; gonidia, 21n; lichens and, 14, 45n; Lyubimenko, 85; mastigotes, acquired by, 48; and mycoplasm, 42; oxygen from, 113; plastids from, xvi, xix, 40–41, 72–73, 102–3, 112, 115, 116–17; Ryzhkov, 107, 112; as schizophytes, 118; term, use of, xxix; *Woronichinia,* 10

Cyanobionta, 118

Cyanophora, 14, 60, 96, 97

Cyanophyceae, 40, 41, 44, 70, 85

cyanophytes, ix, xxix, 116

cyanoses, 51. *See also* endocyanoses

cycads, 75

cyclopeptide liver toxins, 77

Cyclostoma, 76

Cystococcus, 21–22, 53

cytodes, 70–72, 73–74

cytokinesis, 99

Index

Index

Index